T0135220

Studies in Fuzziness and Soft Computing

Volume 407

The series "Studies in Fuzziness and Soft Computing" contains publications on various topics in the area of soft computing, which include fuzzy sets, rough sets, neural networks, evolutionary computation, probabilistic and evidential reasoning, multi-valued logic, and related fields. The publications within "Studies in Fuzziness and Soft Computing" are primarily monographs and edited volumes. They cover significant recent developments in the field, both of a foundational and applicable character. An important feature of the series is its short publication time and world-wide distribution. This permits a rapid and broad dissemination of research results.

Indexed by SCOPUS, DBLP, WTI Frankfurt eG, zbMATH, SCImago.

All books published in the series are submitted for consideration in Web of Science.

More information about this series at http://www.springer.com/series/2941

Mahmut Dirik · Oscar Castillo ·
Fatih Kocamaz

Vision-Based Mobile Robot Control and Path Planning Algorithms in Obstacle Environments Using Type-2 Fuzzy Logic

Mahmut Dirik
Department of Computer Engineering
Şırnak University
Malatya, Turkey

Oscar Castillo
Division of Graduate Studies
Tijuana Institute of Technology
Tijuana, Mexico

Fatih Kocamaz
Department of Computer Engineering
Inonu University
Malatya, Turkey

ISSN 1434-9922 ISSN 1860-0808 (electronic)
Studies in Fuzziness and Soft Computing
ISBN 978-3-030-69249-0 ISBN 978-3-030-69247-6 (eBook)
https://doi.org/10.1007/978-3-030-69247-6

This Springer imprint is published by the registered company Springer Nature Switzerland AG
The registered company address is: Gewerbestrasse 11, 6330 Cham, Switzerland

Preface

The use of mobile robots is becoming increasingly widespread in the leading sectors, especially in industrial areas, such as services, health, defense, and so on. Recent studies and research have focused on systems that enable the mobile robot to move autonomously. In mobile robot systems, it is crucial to draw up an appropriate (cost-effective and safe) path plan to be followed by the robot and to design an excellent controller to characterize the behavior modeling of the robot. Even though robotic control is made more robust with the developing technology, it should be taken into consideration that the costs of robot control increase. The development of low-cost, high-reliability control systems is one of the topics covered in the field of robot control. These issues were taken into consideration when determining the scope of this book.

Within the scope of this book, the path planning for a mobile robot was realized with the external device configuration and control methods that model mobile robot movements that were developed in this planned path. The environment for simulation and real-time applications was developed using LabVIEW software for mobile robot behavior modeling and out off-device configuration. The external eye configuration applied in the scope of the study is preferred since it does not require the use of internal sensors on the robot. The robot, targets, and obstacles detected in the configuration environment by image processing methods were used to create the environment map.

In the generated environment map, a low-cost and safe path was built between the robot and the target without colliding with obstacles using path planning algorithms. For this purpose, path planning methods are determined. Two path plan methods based on fuzzy logic were designed as Type-1 and Type-2. Rule sets for these path planning methods were created. The proposed path planning methods were compared with the existing path planning algorithms in terms of performance and stability. These methods are A*, Random Branching Trees (RRT), RRT + Dijkstra, Bidirectional Random Branching Trees (BRRT), B-RRT + Dijkstra, Probabilistic Road Map (PRM), Artificial Potential Area (APF), and Genetic Algorithm (GA) planning algorithms. In an external eye-type configuration, image-based control approaches, also known as visual servoing, are used. These approaches were used to generate input to the control method according to the global position of the objects on the image obtained from the working environment. To calculate this control input, a distance-based triangular design has been created. With this structure, the robot is controlled according to the distance value of each side of a triangular structure formed between the labels (control points) placed on the robot and the target. Two new methods based on Type-1 and Type-2 fuzzy logic were designed as a control method. Control rule sets were formed and applied to the obtained path plan.

To evaluate the performances of the proposed controller, a comparison was made with Gaussian and decision tree-based controller method specially designed for visual-based servoing in previous studies. Developed path planning and controller algorithms were tested in five different configuration spaces.

In terms of path planning, it is observed that the proposed path planning methods are the best in an average performance in all configuration spaces. Path plan performance was also evaluated by statistical performance metrics in the form of standard deviation, mean, and total error, in particular, the length and execution time of the obtained path plan.

Tests were conducted in a real environment for the controller, which characterizes the robot movements to follow the resulting path plan. The controllers designed according to the test results have been more successful in most of the configuration environments than other methods. For the final evaluation, the results between the simulation and the path created by the robot in the real environment were examined, and it was found that the designed methods were remarkably close to the path plans.

In this work, the mobile robot's go-to goal behavior, obstacle avoidance behavior, and the resulting path plan tracking behavior were successfully modeled in an environment monitored by external eye camera configuration based on a visual servo. According to the results of the study, designed path planning methods, and developed controllers provided significant results that could inspire future studies in this field.

Keywords: Visual-based control, Overhead camera, Mobile robot path planning, Mobile robot path tracking, Type-1/Type-2 fuzzy logic control, Interval type-2 fuzzy inference system (IT2FIS), Distance-based triangle structure (DBTS)

Malatya, Turkey — Mahmut Dirik
Tijuana, Mexico — Oscar Castillo
Malatya, Turkey — Fatih Kocamaz
October 2020

Contents

Abbreviations

ABTS	Angle-Based Triangle Structure
AMR	Autonomous Mobile Robots
APF	Artificial Potential Fields
BRRT	Bidirectional Rapidly Exploring Random Trees
CCM	Cross-correlation matrix
CD	Controller Design
DBTS	Distance-Based Triangle Structure
DTC	Decision Tree Control
ED	Euclidean Distance
FLC	Fuzzy Logic Control
FOU	Footprint of Uncertainty
GA	Genetic Algorithms
GC	Gaussian Control
GPS	Global Positioning System
IBVS	Image-Based Visual Servoing
IT2FIS	Interval Type-2 Fuzzy Inference System
LMF	Lower Membership Function
LUT	Lookup Table
LWA	Left Wheel Angle
LWS	Left Wheel Speed
MBN	Map-Based Navigation
MFs	Membership Function
NRC	Next Range Condition
PBVS	Position-Based Visual Servoing
PRM	Probabilistic Road Map
RGB	Red-Green-Blue
RRT	Rapidly Exploring Random Tree
RWA	Right Wheel Angle
RWS	Right Wheel Speed
SC	Soft Computing

T1F	Fuzzy Logic Type-1
T2F	Fuzzy Logic Type-2
TR	Type-Reducer
TSBCD	Triangle Shape-Based Controller Design
UMF	Upper Membership Function
VBC	Visual-Based Control
VB-CD	Vision-Based Controller Design
VBPP	Vision-Based Path Planning
WMR	Wheeled Mobile Robot

Symbols

$\bar{\mu}_{\tilde{A}}(\boldsymbol{x})$	Upper Membership Function
$\underline{\mu}_{\tilde{A}}(\boldsymbol{x})$	Lower Membership Function
$\underline{f}^{\mathrm{i}}$	Lower Firing Degrees
$\bar{f}^{\mathrm{i}}$	Upper Firing Degrees
σ	Standard Deviation
$\boldsymbol{sf}_i$, $\boldsymbol{sf}_j$	Significance Factor Parameters
$z_n\boldsymbol{p}$, $z_n\boldsymbol{t}$	Normalized Z-Score Values for the Path Length and Execution Time
$\mathbf{V_R}$	Right Wheel Linear Velocity
$\mathbf{V_L}$	Left Wheel Linear Velocity
$\mathbf{V}$	Center Linear Velocity of the Robot
ω	Center Angular (Rotational) Velocity of Left Wheel
L	Track Width of the Robot
$\mathbf{f_G}$	Gaussian function
A_n A_m	Bottom and Upper Internal Angles
Θ	Steering Angle (Turning Angle)
C	Center of Mass of a Mobile Robot

List of Figures

List of Tables

Chapter 1
Introduction

The popularity of autonomous mobile robots has been rapidly increasing due to the needs that arise with the developing technology and improved application areas. The autonomous agents are capable of navigating intelligently anywhere using sensor-actuator control techniques. Mobile robots carry out tasks in various areas; industries, space research, room cleaning, tourist guidance, and entertainment applications without any human intervention [1]. In autonomous system applications, there are common problems that need to be solved, such as navigation, performing a given task, etc. There are several types of research in the literature on this subject. Their underlying philosophy is to design and develop intelligent algorithms or techniques that can control the motion behaviors of mobile agents by enabling them to avoid obstacles in static or dynamic environments. In this scope, it is necessary to know two fundamental parameters. These are sensors that enable the robot to communicate with the outside world and control algorithms that model the motion characteristics.

On the other hand, the development of a satisfactory control algorithm for autonomous mobile robots to perform the navigation task with an appropriate strategy is still a matter of extensive research. Problems such as the cost of hardware like sophisticated processing units, sensors (such as encoders, gyroscope, and accelerometer) used in robot design, and the complexity of the control kinematics should be overcome. Such problems may cause errors in the control process or may increase the cost of mobile robot applications. There are two types of errors in robotic systems: nonsystematic and systematic errors. These errors may cause negative consequences like a collision with obstacles, wrong positioning, etc. Therefore, error minimization is a critical issue in the control process, especially for mobile robots. In robotic systems, developing low-cost systems is also a requirement besides minimizing systematic and unsystematic errors. Based on a behavioral architecture and inspired by creative solution disciplines, it is aimed to develop new control architectures/frameworks in this study.

M. Dirik et al., *Vision-Based Mobile Robot Control and Path Planning Algorithms in Obstacle Environments Using Type-2 Fuzzy Logic*, Studies in Fuzziness and Soft Computing 407, https://doi.org/10.1007/978-3-030-69247-6_1

1.1 Aims and Objectives

The vision-based mobile robot path planning and motion control in the indoor application is the focus of this study. The study includes path planning, avoiding obstacles, following the path, go-to-goal control, localization, and visual-based motion control using the developed control architecture with soft computing and artificial intelligence methods. The proposed vision-based motion control strategy involves three stages. The first stage consists of the overhead camera calibration and the configuration of the working environment. The second stage consists of a path planning strategy using several traditional path planning algorithms (A*, RRT, PRT, GA, Type 1 Fuzzy, APF…) and proposed planning algorithm (IT2FIS). The third stage consists of the path tracking process using previously developed Gauss and Decision Tree control approaches and proposed Type-1 and Type-2 (IT2FIS) controllers. Two kinematic structures are utilized to acquire the input values of controllers. These are Triangle Shape-Based Controller Design (TSBCD), which was previously developed in [2–4] and Distance-Based Triangle Structure (DBTS) that is used for the first time in conducted experiments. Four different control algorithms, Type-1 fuzzy logic (T1F), Type-2 Fuzzy Logic (T2F/IT2FIS), Decision Tree Control (DC), and Gaussian Control (GC) have been used in overall system design. The developed system includes several modules that simplify characterizing the motion control of the robot and ensure that it maintains a safe distance without colliding with any obstacles on the way to the target.

The overall aims of this research are to design and develop an efficient motion control strategy for indoor mobile robot path and tracking systems. Minimizing the complexity of conventional robot control kinematics and reducing systematic and unsystematic errors are additional objectives of this study. Different planning and control algorithms were used in the proposed system. Their performances were compared and evaluated. The controllers having the best results have been used in the path tracking phase and compared Gauss and Decision Tree-based controllers.

1.2 Novelty and Contribution of the Research Work

This book study contributes to the literature in several aspects. These contributions are emphasized as follows:

- Two fuzzy-based path planning algorithms that are Type-1 and Type-2 (IT2FIS) are developed, and their rule tables are explicitly created for a visual-based control system.
- Type-1 and Type-2 fuzzy logic-based controllers are developed and compared with previously developed Gaussian and Decision Tree controllers.
- Color tracking and template matching approaches have been used to improve the efficiency of real-time tracking. Previous studies generally used only color

tracking. Template matching was used for the first time in such an architecture to prevent loss of frame due to the similarity of background color and robot labels.
- By increasing the monitoring performance, the number of frames processed is increased to 30 frames. Frame loss actualizes about 1–2 in 30. It is 2–3 in the previous study in 14–15 frames.
- Distance-Based Triangle Structure (DBTS) is used to compute controller inputs for the first time. Angle Based Triangle Structure (ABTS) is used in previous studies. Angle calculation requires a more complex mathematical process compare to distance calculation. Therefore, the distance-based kinematic approach is designed and utilized.
- MATLAB programming environment is utilized in previous studies. However, because of its high performance compared to MATLAB, LabVIEW Programming environment is fully utilized for this study.
- Experimental results are evaluated with statistical performance metrics (standard deviation, average, total error, and Z-Score) for the visual-based control.
- Color thresholding and Template Matching are used for object detection and tracking.
- A new adaptive threshold computation is designed to track the acquired path plan. It used the distance between wheels as a base input. These base distance and front label distance are used to characterize the path tracking process.
- A simulator and real-world experiment environment are designed to perform visual-based control with eye-out-device configuration space in the LabVIEW programming framework.
- A feature-based navigation technique that uses the Soft Computing (SC) algorithms and Artificial intelligence techniques are integrated into a reactive behavior architecture to improve navigation performance.

1.3 Outline of the Book

The rest of this Book is organized as follows. Chapter 1 provides an overview of the work and sets out the aim and objectives of the study. Chapter 2 includes a review of the relevant literature, and of the background work that forms the foundations of this work. Chapter 3 provides problems and potential solutions related to the development of a vision-based path planning and path tracking architecture. All details about the material and method such as Image processing, Visual Based Control (VBC) system, Path planning, Path tracking, and Kinematic analysis processes are presented in Chap. 4. Chapter 5 focuses on the implementation and evaluation of the proposed system, and analysis of the test results is also presented. Finally, the Conclusion and recommendations for future work are summarized in Chap. 6.

References

1. G. Gürgüze, İ Türkoğlu, Kullanım Alanlarına Göre Robot Sistemlerinin Sınıflandırılması. Fırat Üniversitesi Mühendislik Bilim. Derg. **31**(1), 53–66 (2019)
2. M. Dirik, O. Castillo, A. Kocamaz, Visual-servoing based global path planning using interval type-2 fuzzy logic control. Axioms **8**(2), 58 (2019)
3. M. Dirik, A.F. Kocamaz, E. Dönmez, Vision-based decision tree controller design method sensorless application by using angle knowledge, in *Proceedings of SIU* (2016), pp. 1849–1852
4. E. Dönmez, A.F. Kocamaz, M. Dirik, A vision-based real-time mobile robot controller design based on Gaussian function for indoor environment. Arab. J. Sci. Eng. **43**(12), 7127–7142 (2018)

Chapter 2
Literature Review

In this chapter, we have addressed the literature of the various methods used in related work to vision-based control. It includes information from the literature survey about the mobile robot's current situation, backgrounds, current trends, control architectures, navigation, Visual Based Control (VBC) Studies, Global path planning, Sensor theory, Obstacle avoidance, and Soft computing methods. Besides the studies related to vision-based mobile robot systems, it is also evaluated the proposed methodologies. Especially with the technological innovations in the field of telecommunications, software, and electronic devices, the developments in the field of robotics have shown significant progress in the last decade. Intelligent sensors and actuators are key components that facilitate the development of planning and decision-making units that significantly increase the capabilities of mobile robots. In the future, some issues need to be studied to find an appropriate balance between human-assisted systems and fully autonomous systems and to integrate technological capabilities with social expectations and requirements. Mobile robot applications in an unstructured and unpredictable environment face two primary issues: control architecture and navigation. These problems are detailed in the following sections.

2.1 Mobile Robot Control Architecture

The mobile robot control architecture consists of three consecutive processes. Firstly, it collects information surrounding the robot using sensors. Secondly, it plans the behavior of the robot by producing meaningful commands from this information. Thirdly, actions are taken by using the behavior commands it produces [1, 2]. The control architecture creates the steps necessary to achieve the successful autonomous navigation of a mobile robot [3, 4]. A control architecture consists of traditional artificial intelligence (AI) modules where all sensor readings are combined and where

M. Dirik et al., *Vision-Based Mobile Robot Control and Path Planning Algorithms in Obstacle Environments Using Type-2 Fuzzy Logic*, Studies in Fuzziness and Soft Computing 407, https://doi.org/10.1007/978-3-030-69247-6_2

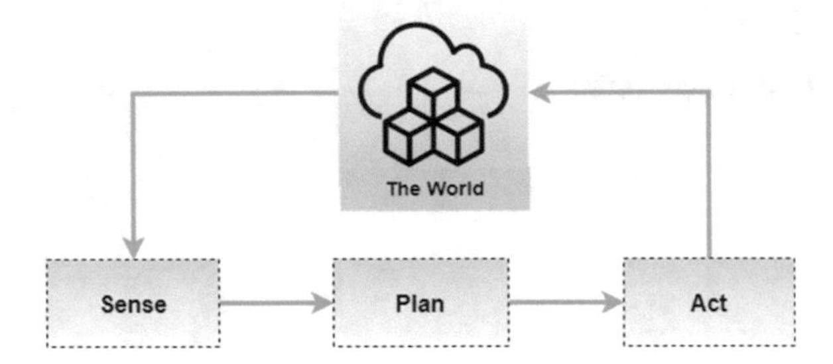

Fig. 2.1 Traditional sense-plan act architecture

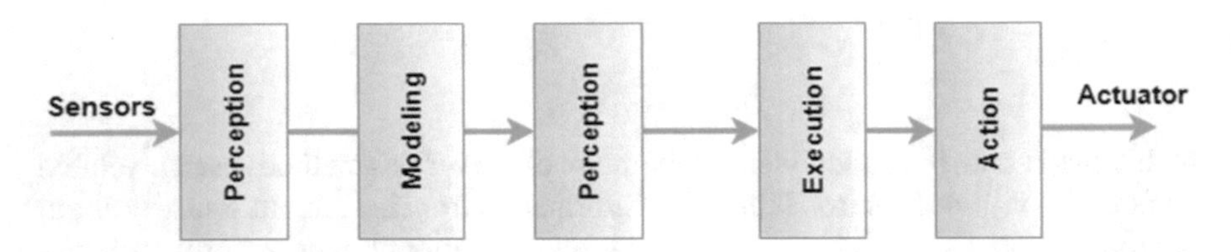

Fig. 2.2 Deliberative architecture

a central planner plans an action and directs the robot accordingly. Figures 2.1, 2.2, and 2.3 illustrate this architecture [4].

The control system uses all sensory processing, modeling, and planning modules together to perform the task or complete the functionality of the behavior [5]. The behavior of a robot, taking a hierarchical approach, perceives its environment in a continuous cycle of planning direction. Then the robot plans its next action based on these feelings, and then take appropriate measures using existing actuators. Thus, at

Fig. 2.3 Reactive/behavior architecture

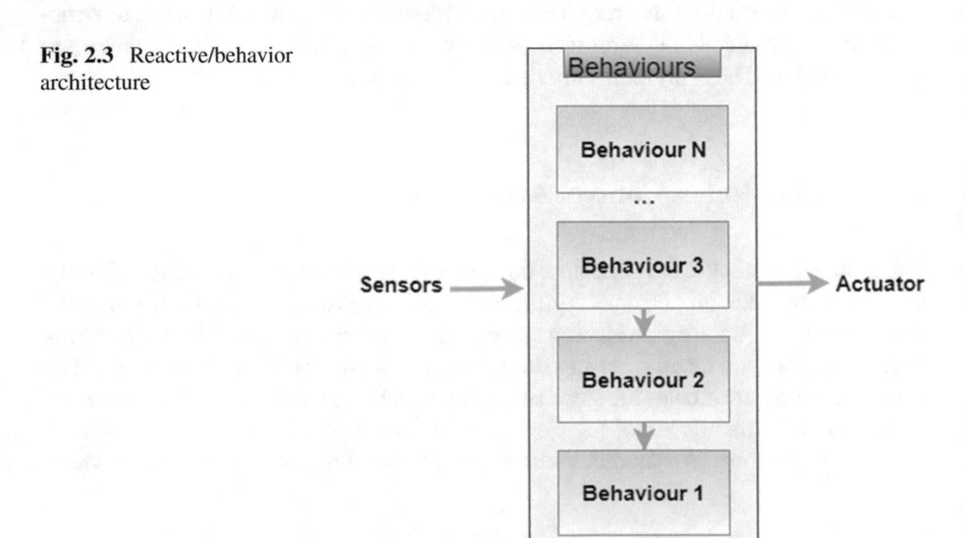

each stage, the robot plans its next action based on the information it has collected so far about the environment. In the realization of complex operations, such functional decomposition can work successfully in a structural environment [6]. Mobile robot control architecture components can be classified into three categories: deliberative, reactive, and hybrid architectures.

2.1.1 Deliberative Architecture

The deliberative or top-down architecture repeats sense, plan, and action steps to plans an optimal trajectory depending on a global world model that is built from sensor data. The deliberative architecture does not rely on the types of complex reasoning processes utilized. Five serial modules, perception, modeling, planning, execution, and action, are principally decomposed into the robot's tasks [5, 6]. These modules do not guarantee the modeling of the robot map and planning a safe path in complex environments. If any of the processes do not work properly in this sequential order, then the entire system may fail. Figure 2.2 illustrates this architecture.

2.1.2 Reactive/Behavior Architecture

Reactive architecture forms the building blocks of more complex behaviors. The information is processed in parallel and not in sequential order. Each parallel data processing module performs a specific task, such as avoiding obstacles or going to the target. The best-known assumption architecture for behavior-based control was introduced by Rodney Brooks [6]. The control system disaggregates a plurality of parallel tasks or behaviors that can directly access sensor data and actuators, as shown in Fig. 2.3.

It is an architectural structure in which the central planner does not need to have comprehensive knowledge. Reactive architecture is divided into two basic classes: subsumption architecture and motor schema [4]. The subsumption architecture was first developed by Brooks [6], and the motor schema was developed by Arkin [7]. Figure 2.4 shows the general graphical representation of a Reactive/behavior architectural structure.

The advantages of this architectural structure are that they can react quickly in dynamic environments, do not need environmental modeling, and are more robust structures because of the different units of behavior.

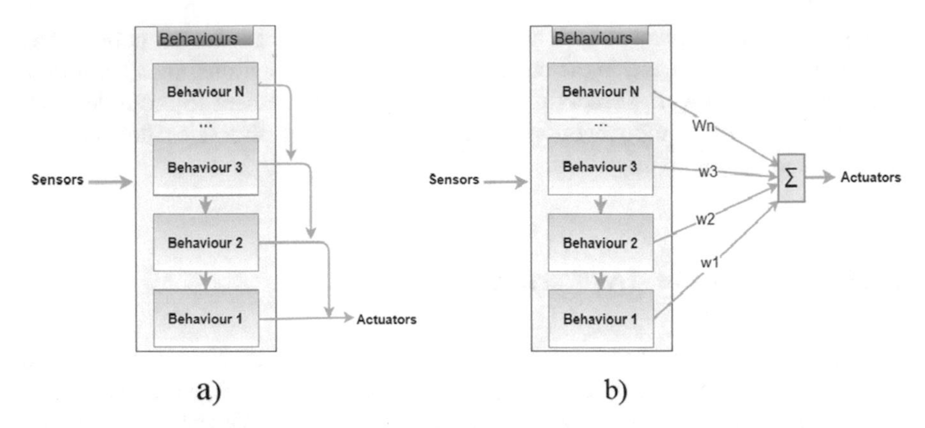

Fig. 2.4. **a** Brooks' subsumption architecture **b** motor schema

2.1.3 Hybrid Architecture

Hybrid control systems are a combination of features from other architectures to create a robust modular control system. Therefore, the selection and integration of architectural features should evaluate each system feature according to the requirements. Hybrid architectures consist of both behavior-based/reactive control and deliberative control architectures [8]. While deliberative is effective in environment modeling and planning, behavior-based control is effective in partial execution of plans and in the rapid response to any unforeseen situation that may arise. Combining the advantages of both deliberative and reactive systems together, it is considered to offer more useful and robust solutions [2, 9]. The hybrid architecture has been illustrated in Fig. 2.5.

The position of the mobile robot is calculated according to a reference starting position based on the wheel speed measured by the encoders. This process is also known as the localization process in mobile systems. This technique is feasible and straightforward in real-time applications. However, situations such as wheel slippage, robot leveling, or physical intervention may cause error accumulation. For minimizing this error, many studies have been carried out [10]. Solutions and suggestions have also been presented using additional sensors such as accelerometer, gyroscope, and compass [11, 12]. The position of the robot has been estimated according to the known starting point in previous studies. This approach is not suitable for indoor applications. However, these solutions increase hardware costs and cannot define the absolute position of the mobile robot. Recently, there have been studies in the literature that predict the current position of a vision-based mobile robot and perform the localization process with lower cost and higher accuracy [13–15].

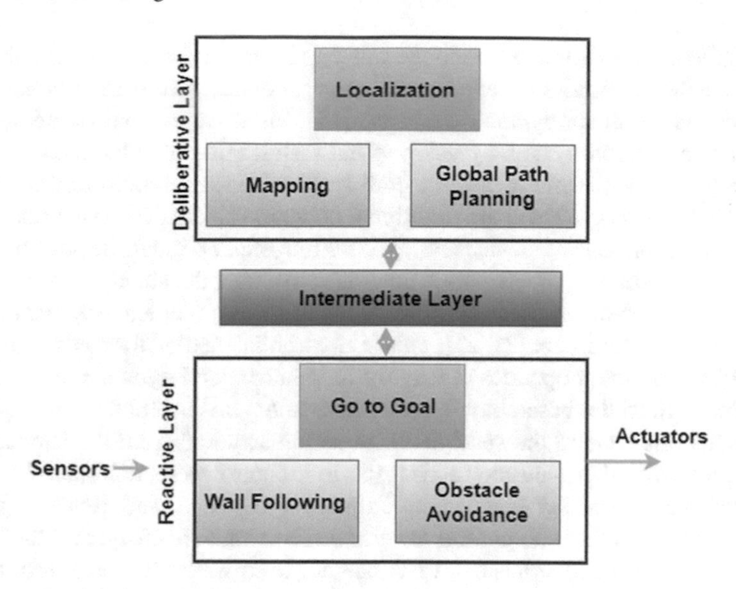

Fig. 2.5 Hybrid architecture

2.2 Mobile Robot Navigation

Navigation is one of the most critical and challenging issues of mobile robot control applications. It includes the determination of an applicable and safe trajectory and all control scenarios that enable the mobile robot to reach the desired target in the predefined trajectory. In an unknown environment, to form the navigation, it is necessary to determine the starting and target positions of the robot and to provide unobstructed path information between start and goal positions. Thus, a mobile robot that can travel independently in a variety of static and dynamic environments can be navigated intelligently anywhere using sensor-actuator control techniques. The navigation problem has been built on the answer to three basic questions. Where am I? Where am I going?

Moreover, how do I get there? The philosophy of all studies in this field is to answer these three basic questions [16]. Various researches have been conducted on mobile robot navigation and avoidance of obstacles [13, 14, 17, 18]. In the next section, detailed information about the studies and applications in the literature related to vision-based navigation concentrating in this work has been given.

2.3 Visual Based Control

Vision-based robot control (also called visual servoing) is one of the powerful and popular research topics in indoor mobile robot navigation that has been studied for decades. It is still an open research area widely used by researchers. In an unknown and unstructured environment, a mobile robot operation needs to cope with dynamic changes in the robot motion environment to navigate successfully to the desired goal while avoiding static or dynamic obstacles [15]. Visual-based control methods aim to manage a dynamic system by using visual features provided by one or multiple cameras [19, 20] with which to acquire both dynamic and static environment information in feedback loops. There are traditional onboard vehicle detection sensors such as sonar, position sensing device (PSD), laser rangefinder, radar. Besides them, the visual-based mobile robot navigation continues to attract the attention of the mobile robot research community because of its ability to acquire detailed dynamic information about the environment [21, 22]. This is an essential method for navigation-based tasks. Visual-servoing operates efficiently in unknown and dynamic environments compared with model-based mobile robot navigation. It is useful for accomplishing the various tasks due to the spacious information acquired from the camera. This information utilized in open-loop and closed-loop control methodologies [23–25]. In this study, the closed-loop vision control algorithm has been used. It is an algorithm where vision detection and control are performed simultaneously, and the control inputs are continuously adjusted. VBC has not been well addressed despite their growing importance in mobile robotics. Optimal path planning, under the view of the camera, can be handled using image-based visual servoing (IBVS) and position-based visual servoing (PBVS) methods [26]. Motion planning methodology is the main difference approach between these methods. The control objective and the control law are directly expressed in the image feature parameter space in IBVS [26]. In PBVS, a 3D camera calibration is required to map the 2D data of the image features to the Cartesian space data. The limitation of both methods combined with developed a $2^{1/2}$ D visual servoing technique which is between the classical position-based and image-based approaches [27, 28]. Vision-based navigation can be examined in three main classes, namely: Map-Based, Map-Building-Based, Mapless approaches [29]. Map-based navigation (MBN) techniques require specific knowledge of the environment, and maps can contain varying degrees of detail between the environment's CAD model and the elements in the environment [30]. In the map building based navigation (MBBN) system first, a mobile robot constructs the environment in a 2D or 3D model using its on-board sensors, then the robot tracks extracted features and computes the optimum path [31, 32]. A mapless navigation system contains all of the navigation that takes place without the knowledge of the environment. Robot navigation is acquired by observing and extracting relevant information about surrounding objects or obstacles [33].

2.4 Mobile Robot Path Planning

Path planning is one of the fundamental topics in the robot control process. It aims to find a safe and short trajectory from the start point to the goal point with obstacle avoidance capability. In path planning, the main problem is to generate a path that allows a robot to move from a starting point to the goal point without colliding any obstacles in configuration space. During the last two decades, a great deal of research focuses on the path planning problem [34–37]. To perform a task with the mobile robot finding a feasible solution in critical applications in real-life, one needs to solve path planning and path tracking problems efficiently [38, 39]. The path tracking problem can be described as the process of guiding and controlling the robot to track the trajectory or to keep up the robot on the generated path. The environment type (static or dynamic) and path-planning algorithm are two important factors in solving the path planning problem. The path planning algorithm can be classified into two categories: global (off-line) or local (on-line) algorithms [40, 41]. Global path planning methods required the environment model (robot map) to be static and completely known. The path-planning problem is categorized into classical and heuristic approaches. There are many algorithms designed for global path planning such as A* [42] which is an extension of the Dijkstra algorithm [43, 44], Genetic algorithm (GA) [45–47], Probabilistic Road Map (PRM) [48], Rapidly Exploring Random Tree (RRT) [49], Bidirectional-RRT (BRRT) [50, 51], Artificial Potential Fields (APF) [52, 53], Fuzzy type 1 and Fuzzy Type 2 path planning algorithm [37, 54]. Many studies have used these soft computing and heuristic techniques to generate an effective solution even in complex environments.

Traditionally, different sensing techniques enable the robot to detect obstacles such as infrared detectors, laser scanner, ultrasonic sensors [55–57]. These sensors may cause systematic and non-systematic errors. Systematic errors generally stem from the encoder, sensor, and physical structure of robot parts. However, unsystematic errors generally stem from outside factors such as sliding, hitting, falling. On the other hand, vision sensors provide low-cost motion control and effective in decreasing errors, as mentioned. They are also useful robotic sensors that allow for non-contact measurement of the environment. The vision system provides information about obstacles simultaneously the position and orientation of a mobile robot in the initial and goal position. The proposed control methods aim to control a dynamic system by utilizing visual features extracted from overhead camera images. The main advantage of the visual servoing [58, 17] is that it requires fewer sensor data, suitable to control multiple robots, internal and external sensors on robots generally are not needed, in terms of scalability; it provides more operating area by increasing imagining devices and so on. The hierarchical classification of the path planning methods is shown in Fig. 2.6.

2.5 Sensor Theory and Obstacle Avoidance

Sensor technology has e considerably in the last decade. This technology is an essential part of the autonomous mobile robot. The mobile robot system gathers information about its environment using different sensors taking measurements and translate the measured information to meaningful data to the control system in which it activates and navigates. A wide range of low-cost sensor systems is available with their unique capabilities and designations that can easily be deployed on robots. A general

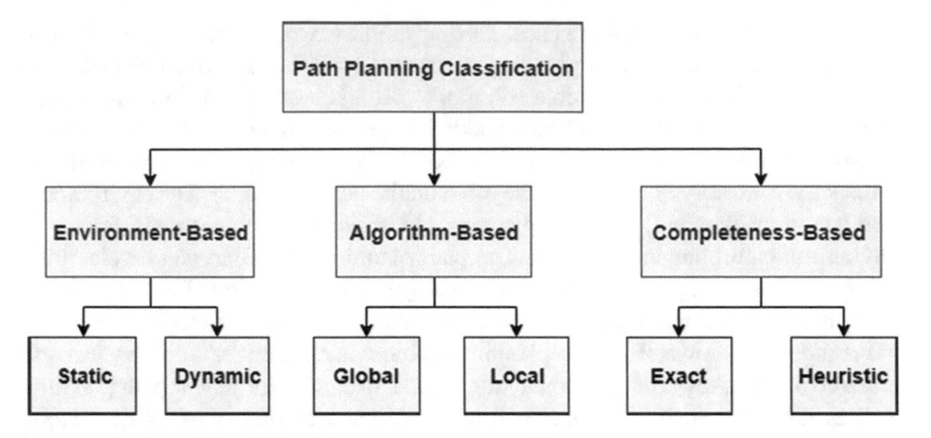

Fig. 2.6 Path planning categories

classification of sensors can be basically into Internal status sensors (proprioceptive) and External status sensors (exteroceptive) sensors groups. Internal status sensors, measure internal values like battery voltages, wheel speed. External status sensors, acquire information from the environment, like a distance from an obstacle, or global position. The sensors collecting information from the real world environment can be classified into active energy emitting, and passive energy receiving sensors, according to their functions based on their interaction with the environment [59]. Depending on the type of measurement sensors can be categorized into Distance sensors (infrared (IR) sensors, Ultrasonic Sensors, Laser Sensors) [60, 61] Positioning sensors (Global Positioning System (GPS)) [62, 63], Ambient sensors (Pyroelectric sensors) [64], and Inertial sensors (accelerometers or gyroscopes) [65].

Vision sensors are one of the considered passive sensors using to model a dynamic or static system providing the most comprehensive information by employing visual features obtained from images provided by the camera. There is an architecture of a visual servoing sensor system broadly implemented in robotic research. This architecture has several significant advantages. First, unlike conventional controllers, the purpose of vision-based control (VBC) is to minimize errors and reduce both software and hardware costs to an acceptable level. In this architecture, there is no need to use onboard sensors. In recent research, image-based visual controllers have been widely used to control autonomous (or self-driving) vehicles [22, 66]. At the same time, real-time robotic systems, multi-tasking robotics, and unmanned aerial vehicles have been developed with the image sensor equipment. A general control model is a control architectural structure that is desirable to have low complexity, low processing time, and high accuracy. The vision-based sensor architecture has a suitable infrastructure for this. All sensor data is used to plan a more reliable path for robots, avoid obstacles, and ensure that a given task is performed with fewer errors.

Obstacle avoidance system is a critical module of autonomous navigation, providing essential information, and protecting mobile robots from collisions while operating in unknown or unstructured static or dynamic environments. Many obstacle avoidance algorithms use active range sensors that furnish direct 3D measurements, such as laser range finders and sonar systems [67, 68]. The general onboard sensors have several drawbacks, such as poor angular resolution (ultrasonic sensor), and high costs (laser sensor). An alternative solution for obstacle avoidance is visual sensors which often provide better resolution range data for obstacle detection and become increasingly popular in robotics [69–74]. Such visual-based systems are dependent on qualitative information techniques also considered in this work. The primary image processing techniques like detecting pixel changes are utilized in image frames to detect the static or dynamic obstacles in real-time. The main advantages of this method are ease of application, efficient, and low cost for real-time applications. The type of obstacles can be classified into two, namely: static and dynamic.

Mobile robot navigation among static obstacles is simple because static obstacle avoidance deals with certain obstacles that never change their shape and position in the environment. The global path planning (off-line) performs in the environments to be static and need complete knowledge about the obstacles. Several algorithms have been proposed to avoid static obstacles [75, 76].

A dynamic obstacle is any moving object that changes its position over time during robot navigation. Control algorithms developed to avoid such obstacles are much more complicated. Different types of static and dynamic obstacles can be placed within the environments for the robot to move without any collision. The overall procedure of obstacle avoidance is shown in Fig. 2.7.

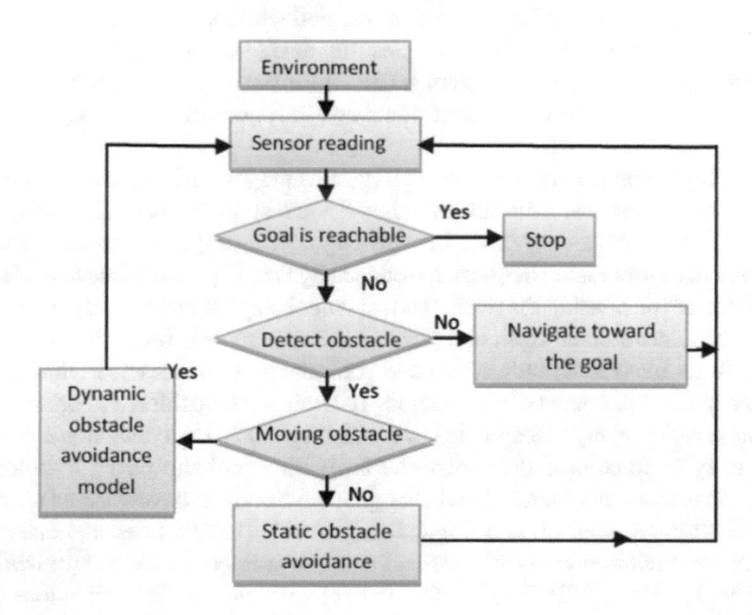

Fig. 2.7 Obstacle avoidance procedure

2.6 Soft Computing Methods in Path Planning

Over the past two decades, the soft computing field has rapidly matured in highly multidisciplinary applications in various domains. According to Professor Lofti Zadeh, soft computing is "an emerging approach to computing, which parallels the remarkable ability of the human mind to reason and learn in an environment of uncertainty and imprecision" [77]. Another definition of soft computing has been provided by Professor Zadeh states, "The guiding principle of soft computing is to exploit the tolerance for imprecision and uncertainty to achieve tractability, robustness, and low solution cost" [78, 79]. Soft Computing (SC) consists of several combinations of computing paradigms, including fuzzy logic, neural networks, and genetic algorithms, which can be used to create robust hybrid intelligent systems [80]. These hybrid architectures include many fields that fall under various categories in Artificial Intelligence. Soft computing techniques provide alternative and more straightforward solutions to mobile robot navigation and obstacle avoidance problem in various environments. The extending of T1F, which is called T2F in this study, can built a powerful hybrid intelligence system that combining with traditional SC techniques, help solve complex control problems [80]. T1F clusters used in conventional fuzzy systems cannot adequately cope with the current uncertainties in intelligent systems. In the evaluation of intelligent systems used in real-world applications based on the soft computation of the computer system in dealing with uncertainties, T2F clusters have been observed to be an essential method with more parameters to manage these uncertainties better. In these studies, the design of intelligent systems using an interval type-2 fuzzy logic system (IT2FIS) for the non-linear control system in which the antecedent or consequent membership functions (MFs) are T2F sets is handled.

Fuzzy logic controllers have compelling advantages such as low cost, ease of control, and designable without knowing the exact mathematical model of the process. They are extensively used in many engineering applications such as mobile robotics, image processing has been introduced by [78, 81]. That is because of simple designable and decreasing the mathematical model complexities. Fuzzy logic can be used in the decentralized form and preferable to the mobile robot than centralized control. In mobile robot applications and path planning, many uncertainties play a vital role in this field may be encountered. To control the position and orientation of the mobile robot, many researchers have utilized a fuzzy logic technique. An intelligent fuzzy logic controller to solve the navigation problem of a non-holonomic mobile robot in an unstructured and changing environment have various uncertainties such as input, control, and linguistics [82, 83]. Uncertainties associated with changing unstructured environments can cause a problem in the determination of MFs. The developed IT2FIS is suitable to deal with real-world applications on the control of a mobile robot [84–87]. This ability is supported by the fact that the third T2F sets dimension and its footprint of uncertainty (FOU) is sufficient as a comparison with T1F sets in modeling on uncertainty. To cope with uncertainties recently, advances made on T2F have been used to develop an intelligent vision system based on IT2FIS for global path planning and path tracking [88]. The lack of fuzzy systems in adapting to changing situations is complemented by combine fuzzy logic with neural networks or genetic algorithms [80]. A hybrid soft calculation method is also the control architecture created in conjunction with the genetic algorithm (GA). There are many studies in the literature about autonomous vehicles using GA based path planning problem in the complex environment [89, 90]. GA-Fuzzy algorithms also have been designed to tuning the best membership function parameters from the fuzzy inference system to optimize the navigation of a mobile robot [91]. In the literature review given above, it was seen that many researchers showed only computer simulation results on mobile robot navigation and avoidance of obstacles based on nature-inspired algorithms. In this work, these algorithms have been implemented in real-time and real robot applications that can effectively solve the mobile robot's navigation and obstacle avoidance problems in static and dynamic environments.

References

1. M.S. Guzel, Mobile robot navigation using a vision based approach, Sch. Mech. Syst. Eng. Newcastle Univ. United Kingdom, Degree Dr. Philos., 2012, 14
2. A. Alsaab, R. Bicker, Behavioral strategy for indoor mobile robot navigation in dynamic environments. Int. J. Eng. Sci. Innov. Technol. **3**(1), 533–542 (2014)
3. S.F. Heidari, Autonomous navigation of a wheeled mobile robot in farm settings, Dr. Philos. Dep. Mech. Eng. Univ. Saskatchewan, Saskatoon, 2014
4. A. Pandey, Mobile robot navigation in static and dynamic environments using various soft computing techniques, Dr. Philos. Dep. Mech. Eng. Natl. Inst. Technol. Rourkela, July 2016, 226
5. R. Siegwart, I. R. Nourbakhsh, D. Scaramuzza, R.C. Arkin, *Introduction to Autonomous Mobile Robots* (MIT Press, 2011)
6. R. Brooks, A robust layered control system for a mobile robot. IEEE J. Robot. Autom. **2**(1), 14–23 (1986)
7. R.C. Arkin, Motor schema—based mobile robot navigation. Int. J. Rob. Res. **8**(4), 92–112 (1989)
8. Y. Yongjie, Z. Qidan, C. Chengtao, Hybrid control architecture of mobile robot based on subsumption architecture, in *International Conference on Mechatronics and Automation* (2006) pp. 2168–2172
9. P. Nattharith, M.S. Guzel, An indoor mobile robot development: a low-cost platform for robotics research, in *International Electrical Engineering Congress (iEECON)* (2014), pp. 1–4
10. K. Nagatani, S. Tachibana, M. Sofne, Y. Tanaka, Improvement of odometry for omnidirectional vehicle using optical flow information, in *IEEE/RSJ International Conference on Intelligent Robots and Systems (IROS 2000) (Cat. No.00CH37113)* (2000), pp. 468–473
11. J. Cobos, L. Pacheco, X. Cufi, D. Caballero, Integrating visual odometry and dead-reckoning for robot localization and obstacle detection, in *IEEE International Conference on Automation, Quality and Testing, Robotics (AQTR)* (2010), pp. 1–6
12. T. Saito, K. Kiuchi, Y. Kuroda, *Mobile robot localization system in frequent GPS-denied situations*, in *2014 IEEE International Conference on Robotics and Automation (ICRA)* (2014), pp. 3944–3949
13. N.J. Cowan, J.D. Weingarten, D.E. Koditschek, Visual servoing via navigation functions. IEEE Trans. Robot. Autom. **18**(4), 521–533 (2002)

14. M. Dirik, A.F. Kocamaz, E. Donmez, Static path planning based on visual servoing via fuzzy logic, in *SIU* (2017), pp. 1–4
15. H. Hadj-Abdelkader, Y. Mezouar, P. Martinet, Path planning for image based control with omnidirectional cameras (2008), pp. 1764–1769
16. S. Khan, M.K. Ahmmed, Where am I? autonomous navigation system of a mobile robot in an unknown environment, in *International Conference on Informatics, Electronics and Vision (ICIEV)* (2016), pp. 56–61
17. F. Bonin-Font, A. Ortiz, G. Oliver, Visual navigation for mobile robots: a survey. J. Intell. Robot. Syst. **53**(3), 263–296 (2008)
18. N. Axel, J. Christian, E. Bayramoglu, O. Rav, *Visual Navigation for Mobile Robots* (Robot Vision, InTech, 2010).
19. F. Chaumette, S. Hutchinson, Visual servo control. I. Basic approaches. IEEE Robot. Autom. Mag. **13**(4), 82–90 (2006)
20. F. Chaumette, S. Hutchinson, *Visual Servoing and Visual Tracking* (Springer Handbook of Robotics. Springer, Berlin Heidelberg, 2008), pp. 563–583
21. S.R. Bista, P.R. Giordano, F. Chaumette, Appearance-based indoor navigation by IBVS using line segments. IEEE Robot. Autom. Lett. **1**(1), 423–430 (2016)
22. Y. Lu, D. Song, Visual navigation using heterogeneous landmarks and unsupervised geometric constraints. IEEE Trans. Robot. **31**(3), 736–749 (2015)
23. Z. Ziaei, R. Oftadeh, J. Mattila, Global path planning with obstacle avoidance for omnidirectional mobile robot using overhead camera, in *IEEE International Conference on Mechatronics and Automation* (2014), pp. 697–704
24. H. Lategahn, C. Stiller, Vision-only localization. IEEE Trans. Intell. Transp. Syst. **15**(3), 1246–1257 (2014)
25. N. Hacene, B. Mendil, Autonomous navigation and obstacle avoidance for a wheeled mobile robots: a hybrid approach. Int. J. Comput. Appl. **81**(7), 34–37 (2013)
26. G.L. Mariottini, G. Oriolo, D. Prattichizzo, Image-based visual servoing for nonholonomic mobile robots using epipolar geometry. IEEE Trans. Robot. **23**(1), 87–100 (2007)
27. A. Assa, F. Janabi-Sharifi, Virtual visual servoing for multicamera pose estimation. IEEE/ASME Trans. Mech. **20**(2), 789–798 (2015)
28. V. Lippiello, B. Siciliano, L. Villani, Eye-in-hand/eye-to-hand multi-camera visual servoing, in *IEEE Conference on Decision and Control* (2005), pp. 5354–5359
29. G.N. Desouza, A.C. Kak, Vision for mobile robot navigation: a survey. IEEE Trans. Pattern Anal. Mach. Intell. **24**(2), 237–267 (2002)
30. J. Levinson, M. Montemerlo, S. Thrun, Map-based precision vehicle localization in urban environments, in *Robotics* (The MIT Press, The MIT Press, 2008).
31. R.J. Heath, C.O. Rock, Inhibition of-ketoacyl-acyl carrier protein synthase III (FabH) by acyl-acyl carrier protein in escherichia coli. J. Biol. Chem. **271**(18), 10996–11000 (1996)
32. A. Sharma, I. Wadhwa, R. Kala, *Monocular camera based object recognition and 3D-localization for robotic grasping*, in *International Conference on Signal Processing, Computing and Control (ISPCC)* (2015) pp. 225–229
33. M.S. Güzel, Autonomous vehicle navigation using vision and mapless strategies: a survey. Adv. Mech. Eng. **5**, 234747 (2013)
34. A. Cherubini, F. Chaumette, G. Oriolo, A position-based visual servoing scheme for following paths with nonholonomic mobile robots, in *IEEE/RSJ International Conference on Intelligent Robots and Systems* (2008), pp. 1648–1654
35. E.A. Elsheikh, M.A. El-Bardini, M.A. Fkirin, Practical design of a path following for a non-holonomic mobile robot based on a decentralized fuzzy logic controller and multiple cameras. Arab. J. Sci. Eng. **41**(8), 3215–3229 (2016)
36. H. Omrane, M.S. Masmoudi, M. Masmoudi, Fuzzy logic based control for autonomous mobile robot navigation. Comput. Intell. Neurosci., 1–10 (2016)
37. J.-Y. Jhang, C.-J. Lin, C.-T. Lin, K.-Y. Young, Navigation control of mobile robots using an interval type-2 fuzzy controller based on dynamic-group particle swarm optimization. Int. J. Control. Autom. Syst. **16**(5), 2446–2457 (2018)

38. J. Han, Y. Seo, Mobile robot path planning with surrounding point set and path improvement. Appl. Soft Comput. **57**, 35–47 (2017)
39. B.K. Patle, D.R.K. Parhi, A. Jagadeesh, S.K. Kashyap, Application of probability to enhance the performance of fuzzy based mobile robot navigation. Appl. Soft Comput. **75**, 265–283 (2019)
40. R. Kala, A. Shukla, R. Tiwari, Robotic path planning in static environment using hierarchical multi-neuron heuristic search and probability based fitness. Neurocomputing **74**(14–15), 2314–2335 (2011)
41. G. Antonelli, S. Chiaverini, G. Fusco, A fuzzy-logic-based approach for mobile robot path tracking. IEEE Trans. Fuzzy Syst. **15**(2), 211–221 (2007)
42. F. Duchoň et al., Path planning with modified a star algorithm for a mobile robot. Procedia Eng. **96**, 59–69 (2014)
43. S.A. Fadzli, S.I. Abdulkadir, M. Makhtar, A.A. Jamal, Robotic indoor path planning using Dijkstra's algorithm with multi-layer dictionaries, in *International Conference on Information Science and Security (ICISS)* (2015), pp. 1–4
44. P. Hart, N. Nilsson, B. Raphael, A formal basis for the heuristic determination of minimum cost paths. IEEE Trans. Syst. Sci. Cybern. **4**(2), 100–107 (1968)
45. C. Lamini, S. Benhlima, A. Elbekri, Genetic algorithm based approach for autonomous mobile robot path planning. Procedia Comput. Sci. **127**, 180–189 (2018)
46. J. Tu, S.X. Yang, Genetic algorithm based path planning for a mobile robot, in *International Conference on Robotics and Automation (Cat. No.03CH37422)* (2003), pp. 1221–1226
47. L. Moreno, J.M. Armingol, S. Garrido, A. De La Escalera, M.A. Salichs, A genetic algorithm for mobile robot localization using ultrasonic sensors. J. Intell. Robot. Syst. Theory Appl. (2002)
48. L.E. Kavraki, P. Svestka, J.-C. Latombe, M.H. Overmars, Probabilistic roadmaps for path planning in high-dimensional configuration spaces. IEEE Trans. Robot. Autom. **12**(4), 566–580 (1996)
49. J. Bruce, M. Veloso, Real-time randomized path planning for robot navigation, in *IEEE/RSJ International Conference on Intelligent Robots and System* (2002), pp. 2383–2388
50. E. Donmez, A. F. Kocamaz, M. Dirik, Bi-RRT path extraction and curve fitting smooth with visual based configuration space mapping, in *International Artificial Intelligence and Data Processing Symposium (IDAP)* (2017), pp. 1–5.
51. R. Sadeghian, S. Shahin, M.T. Masouleh, An experimental study on vision based controlling of a spherical rolling robot, in *Iranian Conference on Intelligent Systems and Signal Processing (ICSPIS)* (2017), pp. 23–27
52. T. Weerakoon, K. Ishii, A.A.F. Nassiraei, An artificial potential field based mobile robot navigation method to prevent from deadlock. J. Artif. Intell. Soft Comput. Res. **5**(3), 189–203 (2015)
53. E. Rimon, D.E. Koditschek, Exact robot navigation using artificial potential functions. IEEE Trans. Robot. Autom. **8**(5), 501–518 (1992)
54. T.W. Liao, A procedure for the generation of interval type-2 membership functions from data. Appl. Soft Comput. **52**, 925–936 (2017)
55. M. Dirik, Collision-free mobile robot navigation using fuzzy logic approach. Int. J. Comput. Appl. **179**, 9 (2018)
56. K. Srinivasan, J. Gu, Multiple sensor fusion in mobile robot localization, in *Canadian Conference on Electrical and Computer Engineering* (2007), pp. 1207–1210
57. A. Shitsukane, W. Cheruiyot, C. Otieno, M. Mvurya, Fuzzy logic sensor fusion for obstacle avoidance mobile robot. IST-Africa Week Conf. 1–8 (2018)
58. A. Remazeilles, F. Chaumette, Image-based robot navigation from an image memory. Rob. Auton. Syst. **55**(4), 345–356 (2007)
59. W.Z. Khan, Y. Xiang, M.Y. Aalsalem, Q. Arshad, Mobile phone sensing systems: a survey. IEEE Commun. Surv. Tutorials **15**(1), 402–427 (2013)
60. G. Benet, F. Blanes, J.E. Simó, P. Pérez, Using infrared sensors for distance measurement in mobile robots. Rob. Auton. Syst. **40**(4), 255–266 (2002)

61. A. Pandey, Mobile robot navigation and obstacle avoidance techniques: a review. Int. Robot. Autom. J. **2**, 3 (2017)
62. C. Cummins, R. Orr, H. O'Connor, C. West, Global Positioning Systems (GPS) and microtechnology sensors in team sports: a systematic review. Sport. Med. **43**(10), 1025–1042 (2013)
63. M. Salvemini, Global positioning system, in *International Encyclopedia of the Social & Behavioral Sciences* (Elsevier, 2015), pp. 174–177
64. D.J. Cook, J.C. Augusto, V.R. Jakkula, Ambient intelligence: Technologies, applications, and opportunities. Pervasive Mob. Comput. **5**(4), 277–298 (2009)
65. B. Barshan, H.F. Durrant-Whyte, Inertial navigation systems for mobile robots. IEEE Trans. Robot. Autom. **11**(3), 328–342 (1995)
66. N. Cao, A.F. Lynch, Inner-outer loop control for quadrotor uavs with input and state constraints. IEEE Trans. Control Syst. Technol. **24**(5), 1797–1804 (2016)
67. A.-C. Hildebrandt, R. Wittmann, D. Wahrmann, A. Ewald, T. Buschmann, Real-time 3D collision avoidance for biped robots, in *IEEE/RSJ International Conference on Intelligent Robots and Systems* (2014), pp. 4184–4190
68. F. Fahimi, C. Nataraj, H. Ashrafiuon, Real-time obstacle avoidance for multiple mobile robots. Robotica **27**(2), 189–198 (2009)
69. J. Gaspar, N. Winters, J. Santos-Victor, Vision-based navigation and environmental representations with an omnidirectional camera. IEEE Trans. Robot. Autom. **16**(6), 890–898 (2000)
70. S. Se, D. Lowe, J. Little, Mobile robot localization and mapping with uncertainty using scale-invariant visual landmarks. Int. J. Rob. Res. **21**(8), 735–758 (2002)
71. L. Carlone, S. Karaman, Attention and anticipation in fast visual-inertial navigation. IEEE Trans. Robot. **35**(1), 1–20 (2019)
72. P. Henry, M. Krainin, E. Herbst, X. Ren, D. Fox, RGB-D mapping: using depth cameras for dense 3D modeling of indoor environments, in *Springer Tracts in Advanced Robotics* (2014), pp. 477–491
73. F. Blochliger, M. Fehr, M. Dymczyk, T. Schneider, R. Siegwart, Topomap: topological mapping and navigation based on visual SLAM maps, in *IEEE International Conference on Robotics and Automation (ICRA)* (2018), pp. 3818–3825
74. M.O.A. Aqel, M.H. Marhaban, M.I. Saripan, N.B. Ismail, Review of visual odometry: types, approaches, challenges, and applications. Springerplus **5**(1), 1897 (2016)
75. A.A. Panchpor, Implementation of path planning algorithms on a mobile robot in dynamic indoor environments, Ph.D. thesis, The University of North Carolina at Charlotte, 2018
76. H. Rezaee, F. Abdollahi, A decentralized cooperative control scheme with obstacle avoidance for a team of mobile robots. IEEE Trans. Ind. Electron. **61**(1), 347–354 (2014)
77. S. Junratanasiri, S. Auephanwiriyakul, N. Theera-Umpon, Navigation system of mobile robot in an uncertain environment using type-2 fuzzy modelling, in *FUZZ-IEEE* (2011), pp. 1171–1178
78. L.A. Zadeh, Soft computing and fuzzy logic. IEEE Softw. **11**(6), 48–56 (1994)
79. L.A. Zadeh, Fuzzy logic, neural networks, and soft computing. Commun. ACM **37**(3), 77–84 (1994)
80. O. Castillo, Introduction to type-2 fuzzy logic control. Stud. Fuzziness Soft Comput. 3–5 (2012)
81. M. Wang, J.N.K. Liu, Fuzzy logic-based real-time robot navigation in unknown environment with dead ends. Rob. Auton. Syst. **56**(7), 625–643 (2008)
82. N.H. Singh, K. Thongam, Mobile robot navigation using fuzzy logic in static environments. Procedia Comput. Sci. **125**, 11–17 (2018)
83. L. Ren, W. Wang, Z. Du, A new fuzzy intelligent obstacle avoidance control strategy for wheeled mobile robot, in *IEEE International Conference on Mechatronics and Automation* (2012), pp. 1732–1737
84. R. Martínez, O. Castillo, L.T. Aguilar, Intelligent control for a perturbed autonomous wheeled mobile robot using type-2 fuzzy logic and genetic algorithms. J. Autom. Mob. Robot. Intell. Syst. **1**(2), 12–22 (2008)

85. O. Castillo, P. Melin, J. Kacprzyk, W. Pedrycz, Type-2 fuzzy logic: theory and applications, in *IEEE International Conference on Granular Computing (GRC)* (2007), pp. 145–145
86. N. Baklouti, R. John, A.M. Alimi, Interval type-2 fuzzy logic control of mobile robots. J. Intell. Learn. Syst. Appl. **04**(04), 291–302 (2012)
87. Q. Liang, J.M. Mendel, Interval type-2 fuzzy logic systems: theory and design. IEEE Trans. Fuzzy Syst. **8**(5), 535–550 (2000)
88. M.A.P. Garcia, O. Montiel, O. Castillo, R. Sepúlveda, P. Melin, Path planning for autonomous mobile robot navigation with ant colony optimization and fuzzy cost function evaluation. Appl. Soft Comput. **9**(3), 1102–1110 (2009)
89. M. Algabri, H. Mathkour, H. Ramdane, M. Alsulaiman, Comparative study of soft computing techniques for mobile robot navigation in an unknown environment. Comput. Human Behav. **50**, 42–56 (2015)
90. AL-Taharwa, A mobile robot path planning using genetic algorithm in static environment. J. Comput. Sci. **4**(4), 341–344 (2008) 341–344.
91. L. Ming, G. Zailin, Y. Shuzi, Mobile robot fuzzy control optimization using genetic algorithm. Artif. Intell. Eng. **10**(4), 293–298 (1996)

Chapter 3
Preliminary Definitions

Path planning, localization, and motion control are common problems in mobile robot control. If the working environment is unstructured and unknown, the navigation problem becomes more difficult. In indoor applications such as households or offices, the system needs the sensor data to overcome the navigation problem, which is represented in the environment. For accomplishing the navigation task using an appropriate strategy, the data is interpreted by the robot's control system. However, research is still underway on the development of a satisfactory control algorithm of conventional built-in hardware sensors and the fact that autonomous vehicles can achieve the desired navigation. All this may be due to several reasons why these sensors may cause systematic and nonsystematic errors and fail to achieve the correct result. The traditional mobile control kinematics and complex mathematical calculations also contain significant functions in this regard. Vision systems have recently been an attraction issue to provide the necessary information about the robot and its environment. This concept is also useful in the design of mobile robots. With vision sensors, progress will be made based on increasing robustness and low cost. Within the framework of these ideas, it is tried to create solutions by developing new vision-based control approaches by creating platforms suitable for mobile robot navigation and methodologies for indoor applications. Figure 3.1 shows the created platform.

This configuration is a designation of the kinematic control structure of vision-based mobile robot navigation in an indoor environment. Using artificial intelligence techniques, it addresses different vision-based aspects for navigation based on obstacle avoidance , localization, and control architecture. The existence of uncertainty in a dynamic and complicated environment makes it challenging to find an optimal path is reasonable. To solve such problems, it is proposed to the combination of the visual system (VS) based on the proposed kinematic control structure for a wheeled mobile robot motion control. The proposed visual control process with a fixed overhead camera minimizes the errors because the robot position is continually tracked and updated according to acquired information from the sequentially

M. Dirik et al., *Vision-Based Mobile Robot Control and Path Planning Algorithms in Obstacle Environments Using Type-2 Fuzzy Logic*, Studies in Fuzziness and Soft Computing 407, https://doi.org/10.1007/978-3-030-69247-6_3

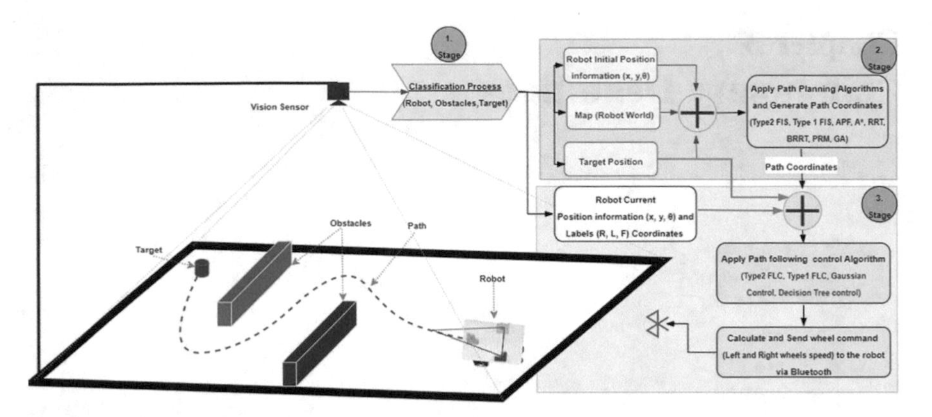

Fig. 3.1 Overall system configuration block diagram

captured images. The robot localization has been measured at each location from these sequential images using template matching and feature extraction methods.

Unlike the traditional global path planning and path tracking algorithms, the proposed algorithms are focused on the implementation of practical real-time model-free algorithms based on the visual servoing system to solve the path planning problem in three stages. First, the proposed algorithm based on visual information extracted from an overhead camera, and the classification process of the position and orientation of the robot, target, and obstacles are handled. Secondly, the initial parameters of the path planning algorithms are determined, and the path coordinates are obtained using these parameters . In this stage, several path planning algorithms have been considered. The third stage handled the path tracking process using the proposed structure to keep up the robot on the generated path. In this work, the mobile robot only performs commands to adjust the speed of the wheels. Because all control processes are applied to an external computer system. The proposed approach is aimed to develop an efficient internal sensor-independent visual-based control method. As a result, it is believed that the developed methods will attract attention in terms of cost, energy efficiency, and robustness.

Chapter 4
Materials and Methods

In this book, a mobile robot path planning and path tracking study was carried out. A *, RTT, RRT + Dijkstra, B-RRT, B-RRT + Dijkstra, PRM, APF, GA, Type-1 Fuzzy Logic, and Type-2 Fuzzy Logic algorithms were implemented, and their performances were compared. The utilized performance measurement parameters used here are path length and execution time of the algorithm. The algorithm with the best performance obtained in the scope of these parameters was used in the second stage, which is the path tracking stage.

Two Fuzzy-based controllers have been developed for the tracking process. The rule sets of the controllers developed are based on the inputs obtained from the distance-based triangular scheme. These control algorithms have been executed on the architectural structure that was proposed in the previous studies [1, 2] and which we proposed as a new control approach to the literature. Many experimental studies have been carried out using the internal angles and edge distances of this architectural structure.

In this book, edge lengths/distances calculated between target/path coordinate and robot control points (robot labels), and these values are used as controller inputs. The experimental results obtained were compared with the values realized by using angle input values in previous studies [1, 2]. In addition to the control algorithms used in the previous study, Type-1 and Type-2 fuzzy logic-based controllers were developed and implemented. A new perspective was introduced to the literature with the Type-1/Type-2 mobile robotic path tracking application made by using this architectural structure. Type-2 Fuzzy logic has been primarily a new control approach developed in this field for both path planning and path tracking (or control). In this study, a system based on only one virtual sensor data has been developed according to the parameters obtained from the kinematic scheme by Type-1 and Type-2 control methods. This way of working reveals the difference in the study from traditional control architectures.

A simulation environment has been developed using LabVIEW software for the applications, as mentioned above. In addition, real-time robot path planning and path

M. Dirik et al., *Vision-Based Mobile Robot Control and Path Planning Algorithms in Obstacle Environments Using Type-2 Fuzzy Logic*, Studies in Fuzziness and Soft Computing 407, https://doi.org/10.1007/978-3-030-69247-6_4

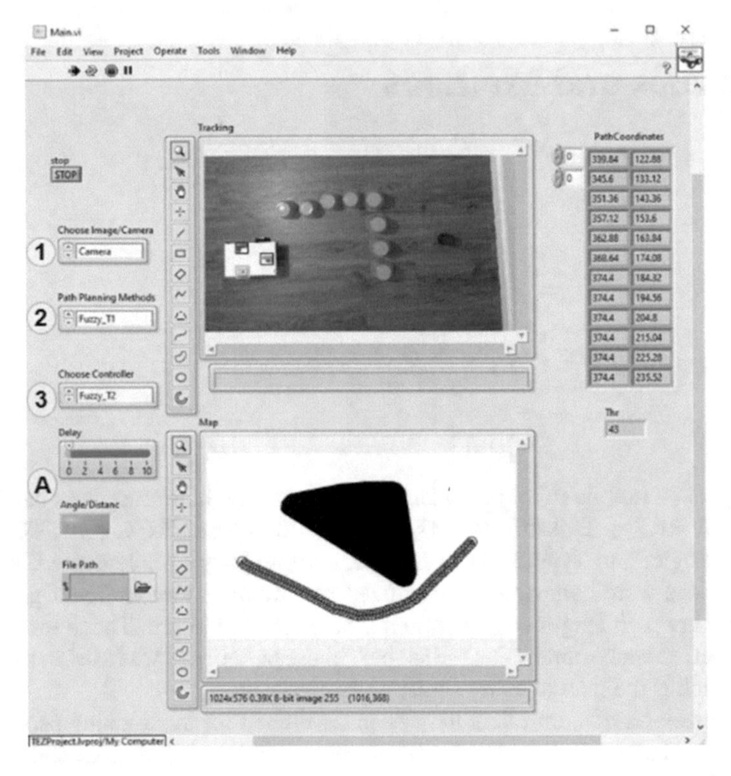

Fig. 4.1 Developed LabVIEW user Interface (front panel)

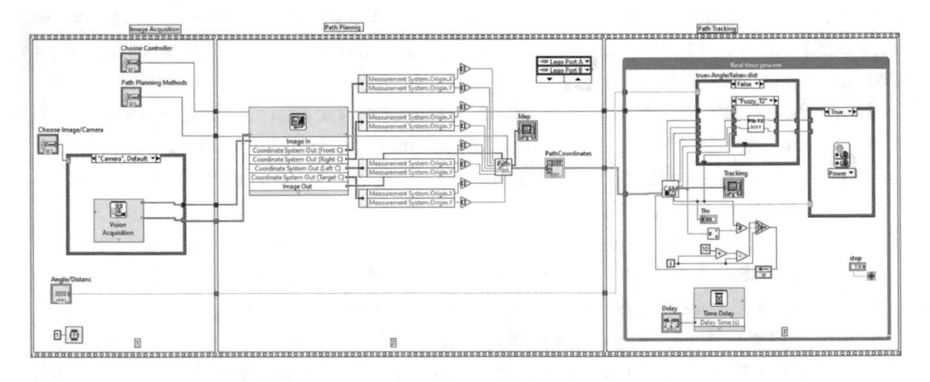

Fig. 4.2 Developed LabVIEW back panel (Code Block)

tracking software have been developed by developing a new user interface with the combination of LabVIEW and Matlab software. The front panel (Fig. 4.1) and back panel (code block) (Fig. 4.2) structure of this developed software are shown below.

This work also includes the design and implementation of a framework aimed to support the development of path planning and path tracking strategies. These are a combination of an intelligent solution using the advantage of a visual servoing system and the IT2FIS system. The proposed control approaches for path planning and path tracking consist of several components. The base infrastructure hardware component has consisted of a mobile robot motion environment, a mobile robot, an overhead camera, and a host computer system. The implemented software component includes both LabVIEW and Matlab image processing tools and control modules. Image acquisition, image processing, object detection and tracking, template matching, parameter acquisition for path planning and path tracking, and calculation of the proposed control structure's parameters are the main modules of the software component.

In the developed LabVIEW user interface, a single image frame can be extracted by selecting a single image frame from Sect. 4.1, or it is provided to monitor the path plan obtained by selecting consecutive image frames from real-time camera images with the controller. In the second part of the interface, the desired path-planning algorithm can be selected. Controller selection is made in Sect. 4.3. The controller developed in this section is compared with the controllers made in previous studies. The Delay section is used to set the simulation speed for path planning. Section 'A' is used to set the distance or angle value as input. In the middle section of the interface, screens are showing both the real environment and the map obtained from the real working environment. On the right side, the coordinates of the obtained path plan are given. "STOP" button is also added to the developed interface to stop the system against possible violations like malfunctions, general safety.

The concept of the proposed methods, especially, is using virtual inputs that are entirely generated from image information. The extracted visual information that is interesting objects under the visibility of the camera has enabled generating the desired path. The proposed algorithm is divided into three stages (see Fig. 4.2).

The first stage consists of image acquisition, the second stage is path planning, and the third stage is designed as a position controller based on the triangle shape based kinematic control structure using proposed control algorithms (Decision tree, Gaussian, Type 1 and Type 2 (IT2FIS) fuzzy logic). Gaussian and Decision tree-based controllers used in previous studies [3] were combined with the Type 1/Type 2 fuzzy logic controller. Each stage is tested separately, and then the overall stages are integrated and tested in various cases to test the validity of the system.

The proposed control methods aim to control a dynamic system by utilizing visual features extracted from the visual servoing system. The main advantages of the visual servoing are that it requires fewer sensor data, suitable to control multiple robots, internal and external sensors on robots generally are not needed, in terms of scalability; it provides more operating areas by increasing imagining devices and so on.

Modern position control environments require a controller to eliminate parameter instabilities and system uncertainties. Soft computing methods like Fuzzy logic is one of these controllers [4, 5]. These controllers have powerful advantages such as low cost, ease of control, and designable without knowing the exact mathematical model of the process. Because of simple designable and decreasing the mathematical model complexities, fuzzy logic can be used in the decentralized form and preferable

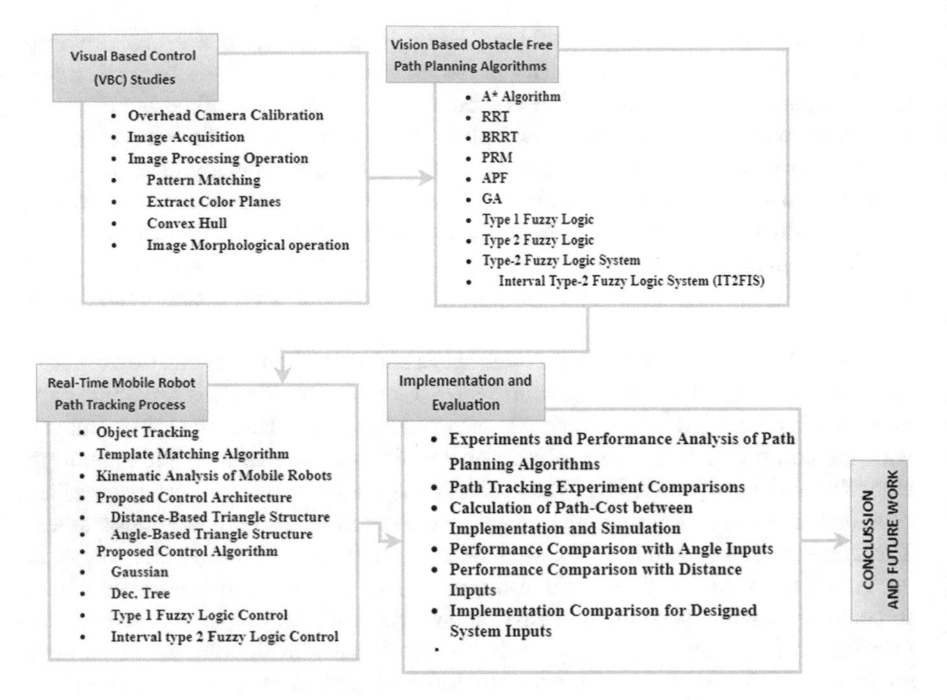

Fig. 4.3 Outline of the research work

to the mobile robot than centralized control. In mobile robot applications and path planning, many uncertainties may be encountered. These are uncertainties of inputs, uncertainties of control action, and linguistic uncertainties. Uncertainties associated with changing unstructured environments and this cause problem in the determination of membership functions. IT2FIS techniques are suitable to deal with these uncertainties [4, 6]. This ability is supported by the fact that the third T2F sets dimension and its footprint of uncertainty is sufficient as a comparison with T1F sets in modeling on uncertainty. IT2FIS is suitable for real-world applications on the control of mobile robots [7, 8]. To cope with uncertainties, we use recently advances made on T2F, which is to develop an intelligent vision system based IT2FIS for global path planning and path tracking. To demonstrate the effectiveness and validity of the proposed system, real-time applications have performed, and the data obtained has shown by graphical representation and interpretation. Figure 4.3 illustrates the stage of the developed system. Details of all steps listed here are given in the following sections.

4.1 Visual Based Control (VBC)

The basics of engineering and research are measurement. Vision development-based control approaches is an alternative measurement unit, which is used to analyze objects in an image. For processing acquired images, hundreds of functions may have been used. In this work, it is used as a general camera as an image acquisition hardware that is connected to the computer through a standard communication interface, as most computers are supplied as USB as standard configurations. For real-time acquisition and image analysis, a USB camera (link) was used with methods using the LabVIEW Vision Development module from National Instruments to apply to image analysis and machine vision.

4.1.1 Overhead Camera Calibration

Camera selection can radically change the image processing needed depending on the application to be made and usually saves processing time during processing. A suitable camera, lens, and lighting arrangement can be convenient to develop the solution. The features of the overhead camera used are as follows: The camera used in the experiments was hung vertically and placed 180 cm above the ground. The camera has a resolution of 3.2 MP, and the lens has a focal length of 0.3 mm and is used in SVGA (Super VGA: 800 × 600; 4:3). Communication with the camera is implemented through the USB 3.0 interface. Experiments have been performed on Intel i7-5500U 2.40 GHz CPU with 16 GB Memory computer.

The position control of the mobile robot is done through visual feedback. It is necessary to convert the 3D Cartesian coordinate of the image to the corresponding pixel coordinates of 2D in the image plane. For this reason, the camera calibration needs to be calibrated appropriately. Distortion problems can be found objectively in every image regardless of the quality of the camera and lens; however, causes such as zooming, complex zooming in lens cameras, and lens structure are more serious, which can further damage image quality. Generally, camera distortion models can be grouped into a radial distortion model and a global distortion model. Radial distortion is caused by defects in the radial curvature of the lens elements, and tangential deformation results from the collinearity of the optical centers of the lens elements. Therefore, distortion correction in the radial model may be sufficient for a fixed lens only in some cases. The following equations are used to solve this problem.

$$x_{corrected} = x\left(1 + k_1 r^2 + k_2 r^4 + k_3 r^6\right) \tag{4.1}$$

$$y_{corrected} = y\left(1 + k_1 r^2 + k_2 r^4 + k_3 r^6\right) \tag{4.2}$$

Therefore, the global distortion model is determined by adding a tangential component to the radial distortion as the Eqs. (4.3) and (4.4).

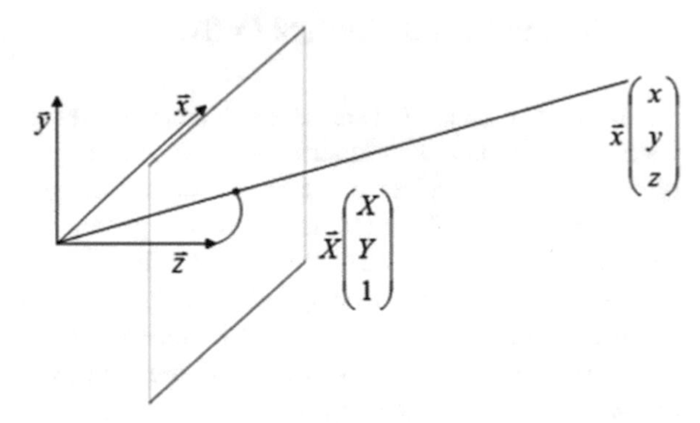

Fig. 4.4 General perspective projection model of a camera

$$\hat{x} = x\left(1 + k_1 r^2 + k_2 r^4\right) + \left(2p_1 \text{xy} + p_2\left(r^2 + 2x^2\right)\right) \tag{4.3}$$

$$\hat{y} = y\left(1 + k_1 r^2 + k_2 r^4\right) + \left(2p_2 \text{xy} + p_1\left(r^2 + 2y^2\right)\right) \tag{4.4}$$

In these equations, the $(\hat{x}, \hat{y})$ pair is the ideal 2D image point without distortion, (x, y) is the distorted point, k_1, k_2, p_1, p_2 are the distortion coefficient with respect to the radial and tangential distortion [9]. A general perspective projection model of a camera is given in Fig. 4.4. The velocity screw of the related frame is given in (4.5).

$$F_s(o, \vec{x}, \vec{y}, \vec{z}) \rightarrow T = (v(o), w) \tag{4.5}$$

$$\overrightarrow{X} = \frac{1}{2}\vec{x} \tag{4.6}$$

In this equation, $v(o)$ is translational velocity, and w is rotational velocity. By assuming the focal length of the camera is equal to '1' then a point with $\vec{x} = (x, y, z)^T$ coordinates are projected to a plane on the image as a point with $\overrightarrow{X} = (X, Y, 1)^T$ by using Eq. (4.6).

4.1.2 Image Acquisition

This section introduces how to acquire and display images. At this stage, the working environment map or image of the configuration area is obtained with an overhead camera configuration. The developed LabVIEW front panel has a button to select the selection of a single image or video frame stream (see Fig. 4.1). Several vision-acquisition-interface methods have been developed, including camera link, USB, IEEE 1394, and GigE to acquire an image from a camera. The advantage of USB cameras is that a USB compatible camera is used because of their relatively inexpensive and 3.0 standard, as well as the speed or bandwidth they reach to support data requirements. Before acquiring, analyzing, and processing images, it is necessary to

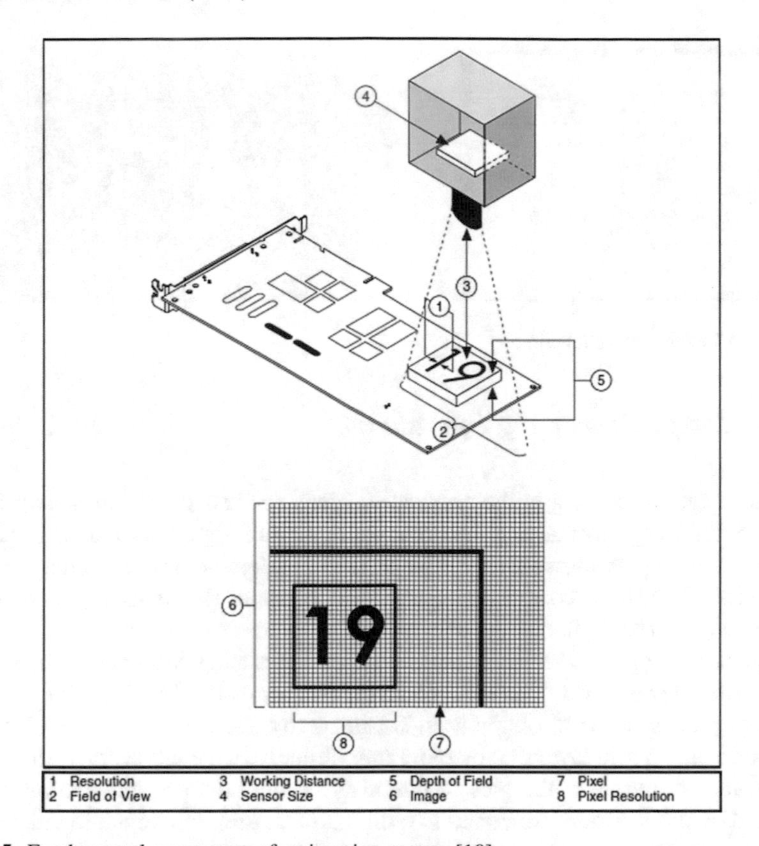

Fig. 4.5 Fundamental parameters of an imaging system [10]

examine and install the imaging system and its components. Field of view, working distance, resolution, and depth of field and sensor size are the factors that make up the imaging system. Figure 4.5 presents these concepts.

The LabVIEW NI Vision Acquisition software module is used for image acquisition, processing, viewing, and recording. Two Vision Express functions are used in the Vision Express palette to acquire and process the image. These are Vision Acquisition Express and Vision Assistant Express. Vision Acquisition Express can receive images from the camera or read the image file using the NI-IMAQ or NI-IMAQdx functions. The Vision Assistant Express function can automate the creation of image processing tasks in the LabVIEW environment [10]. As a result of using Express VI, the block diagram, and Image Out display on the front panel is shown in Fig. 4.6. This is a bird's eye view of the working environment obtained with the camera placed vertically to the floor.

After obtaining the image, the Vision Assistant Express function was used to process this image for the intended purpose. That is used to provide information about the positions of objects (such as obstacles, robot position) in the mobile robot's movement environment. Detailed information about this section is given in the following sections.

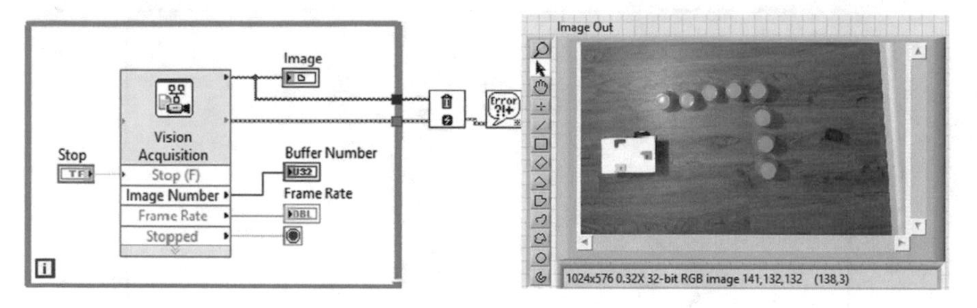

Fig. 4.6 VI for image acquisition

4.1.3 Image Processing Operation

Digital image processing is the application steps for converting an analog image into a digital image format and then processing it with digital computers for image enhancement, repair, classification, compression, understanding, and interpretation. The definition of this process is considered a computer that is used to modify the measured or stored digital image data in an electronic environment.

Digital image processing or computer vision processing steps can be handled in three stages: lower level, intermediate level, and upper-level vision. The first step in image processing is to obtain a digital image by acquiring the image from the real world onto a film layer or a memory unit through image receivers. If the picture sensor does not convert the picture directly to a digital form, the analog picture obtained by the sensor is converted to a digital form utilizing an analog/digital (A/D) converter. Before using the resulting digital image, some pre-processing such as image enhancement, image restoration, and image compression are applied to achieve a more successful result.

Images from the cameras are color or grayscale images. The acquired color image is first converted to a grayscale image and then converted to a binary image. In this transformation process, various morphological functions are applied to remove unwanted objects effectively and to modify the objects of interest for a more accurate measurement. These process steps and conversion methods will be discussed in detail later using LabVIEW Vision Express.

4.1.3.1 Pattern Matching

The pattern matching method is the first image processing method we use in our application. Pattern matching finds regions of a grayscale image that match a predetermined pattern, regardless of lighting, blur, noise, pattern shift, or pattern rotation. Pattern matching algorithms are some essential functions used in various applications in image processing, which has an essential place in machine vision. The template can be defined in one of two ways: template can be defined by drawing an ROI (region of interest) over an existing image, or a template can previously exist as

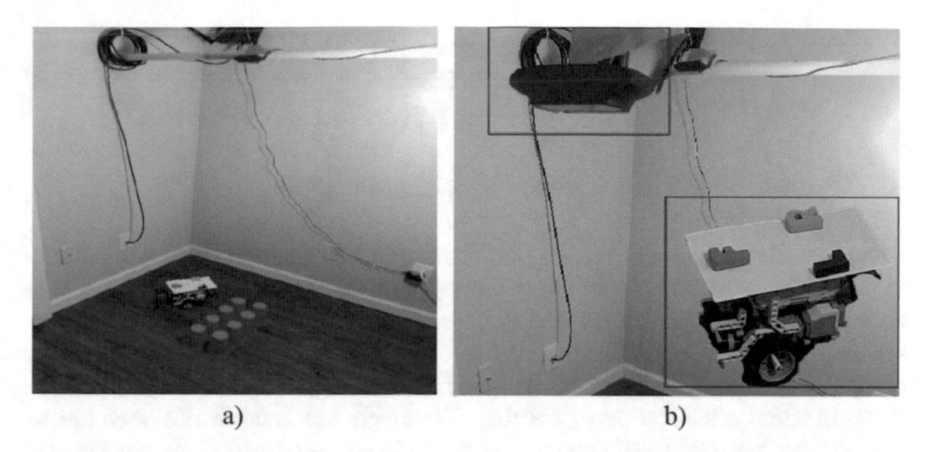

Fig. 4.7 **a** Test environment image, **b** Close-up of the top camera and mobile robot

an image file. Traditional pattern matching techniques include normalized cross-correlation, pyramidal matching, and scale-invariant matching. In this system, we use pattern matching to determine the position of the tags located on the robot in the environment of the overhead camera display, as illustrated in Fig. 4.7.

(A) Cross-Correlation

Conceptually, the template hovered over the source image, and the density values for each corresponding pixel are multiplied separately. All are summed to produce a single correlation value, so the process is repeated until the entire source image is covered, and a matrix of correlation values is created. The standard method used to find a template in an image is the Normalized cross-correlation technique. The correlation process is time-consuming because it is based on a series of multiplication operations. However, new technologies have been developed, such as MMX, which allows performing parallel multiplication and reduce overall computation time. The correlation value matrix, which is most similar to the pattern in the source image, is evaluated as the template with the highest value. To mathematically explain the concept of cross-correlation, consider a source image matrix f (x, y) of size MxN, and a template matrix w (x, y) of size KxL where $K \leq M$ and $L \leq N$. The cross-correlation matrix (CCM) between source and template image at a point (i, j) is performed using the following equation if images are normalized.

$$\mathrm{CCM}_{i,j} = \sum_{x=0}^{L-1} \sum_{y=0}^{K-1} (\mathrm{Template}_{(x,y)} \left(\mathrm{Source}_{(i+x,i+y)} \right) \tag{4.7}$$

In this equation, $i = 0, 1 \ldots M-1$, $j = 0, 1 \ldots N-1$, and the summation is taken over the region in the image where template and source image overlap.

If the images are not normalized, the following formula is used to normalize each item in the corresponding images.

$$\text{CCM}_{\text{i,j}} = \frac{\sum_{x=0}^{L-1}\sum_{y=0}^{K-1}\left(\text{Template}_{(x,y)}\left(\text{Source}_{(i+x,i+y)}\right)\right.}{\sqrt{\left(\sum_{x=0}^{L-1}\sum_{y=0}^{K-1}\left(\text{Template}_{(x,y)}\right)^2\right)\left(\sum_{x=0}^{L-1}\sum_{y=0}^{K-1}\left(\text{Source}_{(x,y)}\right)^2\right)}} \quad (4.8)$$

(B) **Scale-Invariant and Rotation Invariant Matching**

Cross-correlation may not produce the desired result when mapping objects of a different size or rotated source image in the template. This method is considered to be defective in this regard. The problem of image analysis needs to be overcome by re-scanning the template on the source image using different rotations and dimensions (variances in both x and y). For scale-invariant matching, it is necessary to perform the correlation after scaling or resizing. The operation done in this way can be time-consuming. Only the technique of possible scanning variations can be used to accelerate variable matching and accelerate rotation. If the rotation attribute is unknown, searching for the best match requires full template returns. If the size of the template does not change, and there is no spatial distortion, it is not necessary to scan for size scanning. Similarly, if the part is repeatedly placed in the same orientation, it is not necessary to re-scan the source image using a different angle range for rotation variance. A standard method of performing cross-correlation is by rotating to a new angle after placing a small coordinate pointer in the object so that the template can be easily found and predicted in terms of rotation.

(C) **Pyramidal Matching**

In this method, both the image and the template are subdivided into smaller spatial resolutions. The image and template can be reduced to one-quarter of their original size. Reduces both the source image and the template to smaller spatial resolutions, reducing the amount of data to be searched by up to 75%. Matching is performed first on reduced images, making matching faster [10]. Therefore, only areas with very high matching results are considered as the matching fields in the original image.

In this application, the existing template images placed on the mobile robot, as shown in the configuration environment in Fig. 4.8, are used. Robot localization and position information were obtained using these templates (R, L, and F). Each template is uniquely identified by an onboard geometric pattern, which allows the camera to track their location and heading across the testbed individually. Information such as mobile robot start position, start angle relative to the target, and coordinates were obtained from these templates. Thus, some of the input parameters required for path planning algorithms were obtained. The pattern recognition and robot localization then used as reference points for all further measurements in the path planning process.

The position information of the reference images obtained after the template processing operation is given in Fig. 4.9.

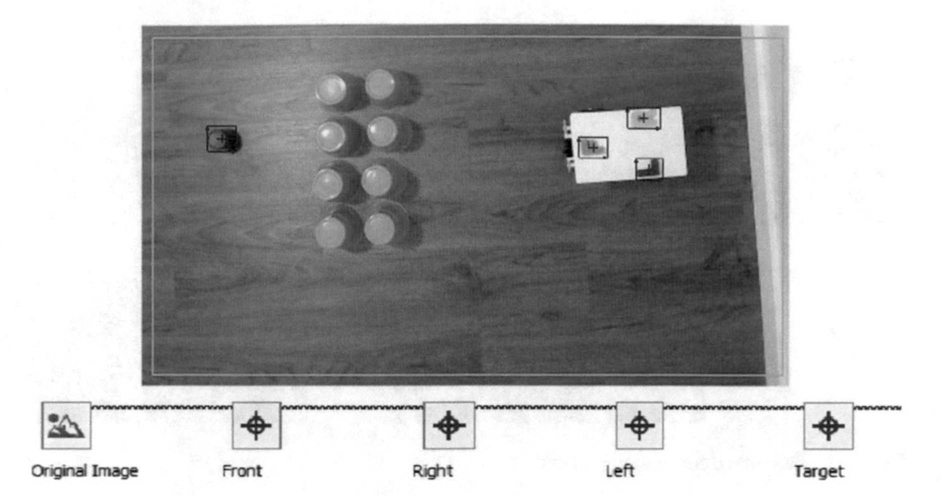

Fig. 4.8 LabVIEW VI implementation of template matching system

Results Viewer

Step Name	Step Type	Result Name	Value	Unit
		Match 1.X Position (Pix.)	705.00	pixels
		Match 1.Y Position (Pix.)	196.00	pixels
		Match 1.Angle (degrees)	0.00	
		Match 1.Score	985.63	
Right	Color Pattern Matching			
		# Matches	1	
		Match 1.X Position (Pix.)	786.00	pixels
		Match 1.Y Position (Pix.)	149.00	pixels
		Match 1.Angle (degrees)	0.00	
		Match 1.Score	823.93	
Left	Color Pattern Matching			
		# Matches	1	
		Match 1.X Position (Pix.)	794.00	pixels
		Match 1.Y Position (Pix.)	231.00	pixels
		Match 1.Angle (degrees)	0.00	
		Match 1.Score	973.23	
Target	Color Pattern Matching			
		# Matches	1	
		Match 1.X Position (Pix.)	129.00	pixels
		Match 1.Y Position (Pix.)	182.00	pixels
		Match 1.Angle (degrees)	0.00	
		Match 1.Score	981.39	

Fig. 4.9 Initial position information of the resulting templates

4.1.3.2 Extracting Color Planes

To analyze any particle in image processing, the color image to be processed must be converted to a gray image and, if necessary, a binary image. Since the camera we use in this application is a color camera and the objects in the camera field of view have different colors and structures, it has been continued the process with converting the color to a gray image and then to a binary image. The grayscale image is converted to a binary image by a threshold operation. If it is necessary to match a color image

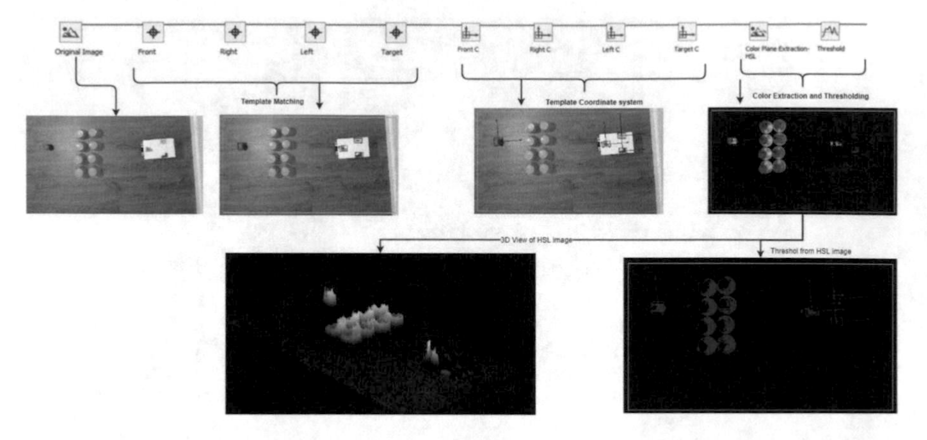

Fig. 4.10 The image conversion process

instead of a gray image, thresholds should be specified for each color plane (red, green, blue planes or hue, saturation, luminance). Some of the image processing steps we have handled for this application are presented in Fig. 4.10.

The template matching process was applied to the color image, as shown in this figure. Then the coordinate system of the received templates was obtained. A threshold was applied to obtain a gray image from a color image followed by a binary image. Here, HSL (color tone and brightness) color space is used because it gives better results against brightness and light change. To obtain an accurate representation of the boundaries of the object, the proper selection of a threshold value is essential for image analysis. The choice of the threshold value, which is for 8-bit gray images are 0 to 255, is a critical step because if the threshold value is not selected correctly, the converted binary image does not display object properties correctly. LabVIEW vision assistant module can be used to adjust the threshold range slider and threshold value easily. As the threshold value changes, the binary image pixels in the converted binary image are displayed in red as binary image pixels 1 and the background color as black. That makes it easier to select objects of interest from the background. The convex Hull method was applied after image thresholding. This method is applied to eliminate the distortion in the image and to eliminate the problems that may cause local minima and collisions. Details of this method are provided in the next section.

4.1.3.3 Convex Hull

The convex hull method is essentially a generalized set of dots or envelopes that contain all the points in the set. The convex hull function is used to combine this set of points as a single particle. The general calculation formula of convex hull method is given in Eq. (4.9).

$$\text{Conv}(S) = \left\{ \sum_{i=1}^{|S|} a_i x_i \,|\, (\forall i : a_i \geq 0) \wedge \sum_{i=1}^{|S|} a_i = 1 \right\} \tag{4.9}$$

where S is the point set, x_i is the array of points, and a_i is the assigned weight or coefficient such that the sum of all x points multiplied by their respective weight is equal to one? In the literature, there are also different calculations for the set of points connected depending on the point-to-point angle [11, 12].

The first reason for using the convex hull technique in this study is to eliminate the local minimum problem when performing mobile robot path planning. The second reason is to develop a high-performance path-planning algorithm without unnecessarily making too many routes and taking a long time.

In Fig. 4.13, the convex hull application and other processes are shown together. In this figure, images (red dots) that treat individual pixels as points of interest have formed a perimeter around the convex body or a set of pixels. Preparing the binary image for the morphological skeleton is the main application of the convex hull method. Here, the positions of the obstacles were determined, and the starting position information of the mobile robot's labels (L, R, and F) and the target were obtained. After the convex hull process, the image must be enlarged or magnified at the specified frame rate to avoid friction and hitting obstacles during the robot path tracking. These operations are discussed below.

4.1.3.4 Image Morphological Operation

After converting the resulting image to a binary image, some morphological functions may be required. The primary reason for using these functions is to remove unwanted particles, isolate bound particles, or improve the dual representation of the particles. Objects or particles are consistent groups of pixels with the same properties, therefore adding or removing pixels to the boundaries of these groups of pixels refers to the morphology function. The rectangular or hexagonal frames surrounding the pixel (see Fig. 4.11) form the configuration elements for calculating the new pixel values. Depending on the connection structure of neighboring pixels, four pixels or eight pixels are considered.

The symbols above are used to show or adjust the shape and connection of the structural element created by the IMAQ Vision Builder. It is formulated as follows according to the number of neighbor relations (4 or 8 connectivity) to be used in the calculation of a pixel value.

$$s_0' = f(s_2, s_4, s_5, s_7) \tag{4.10}$$

$$s_0' = f(s_1, s_2, s_3, s_4, s_5, s_6, s_7, s_8) \tag{4.11}$$

In IMAQ Vision Builder, it is also possible to define larger components such as 5 × 5 and 7 × 7.

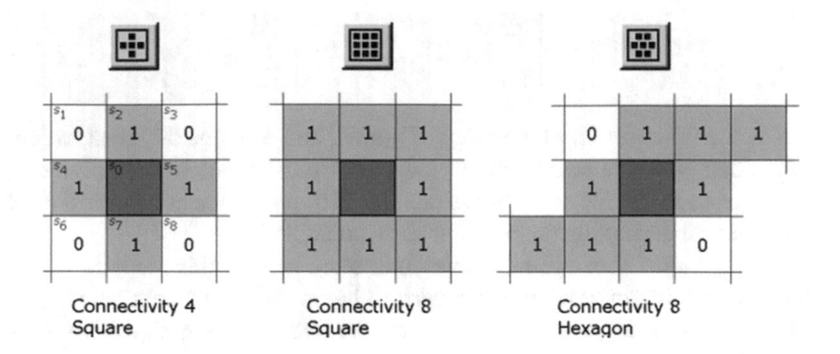

Fig. 4.11 Examples of structuring elements [13]

(A) Erosion and Dilation

Erosion and Dilation are two fundamental algorithms that may be needed in many of the morphological in image processing. The erosion function allows removing pixels from the boundary of particles or objects. In other words, erode reduces the size of all objects by eroding the outline of the object and can effectively eliminate small objects. Dilation allows us to increases the size of objects to expand, and it is provided to fill holes and connect disjoint objects [14]. An example of the morphological operations of erosion and dilation with the used structural elements is shown in Fig. 4.12.

(B) Remove Small Objects

As shown in Fig. 4.13, after the convex hull function is executed, 'Remove Small Objects' function is applied to eliminate unwanted particles. The removal of these unwanted small objects is accomplished through LabVIEW using IMAQ Remove Particle VI in the basic morphology LabVIEW palette [15]. It shows that many small particles or noises are effectively removed.

(C) Equalize and Inverse

The Equalize function called 'IMAQ Equalize' is a lookup table (LUT) operation that the LUT is computed based on the content of the image where the function is applied. Apply LUT transformations to highlight input grayscale values in the

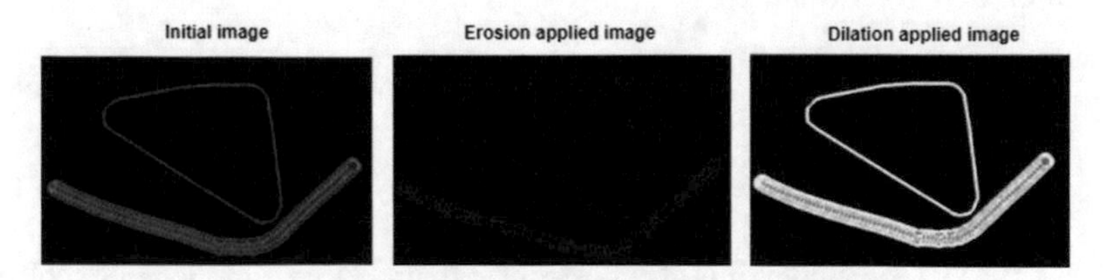

Fig. 4.12 Application of morphological operations: original image, erosion, dilatation

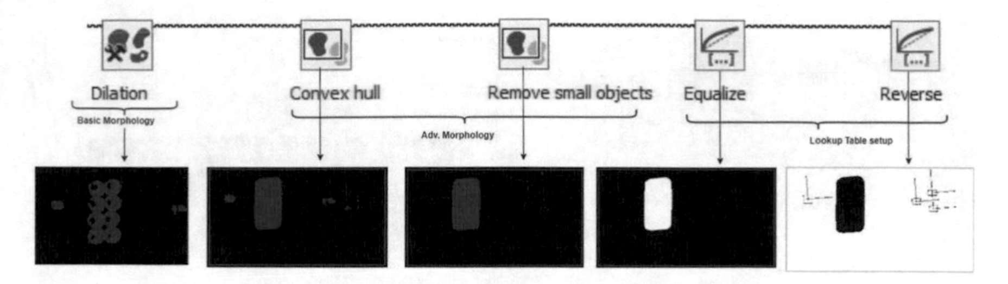

Fig. 4.13 Image morphological operation

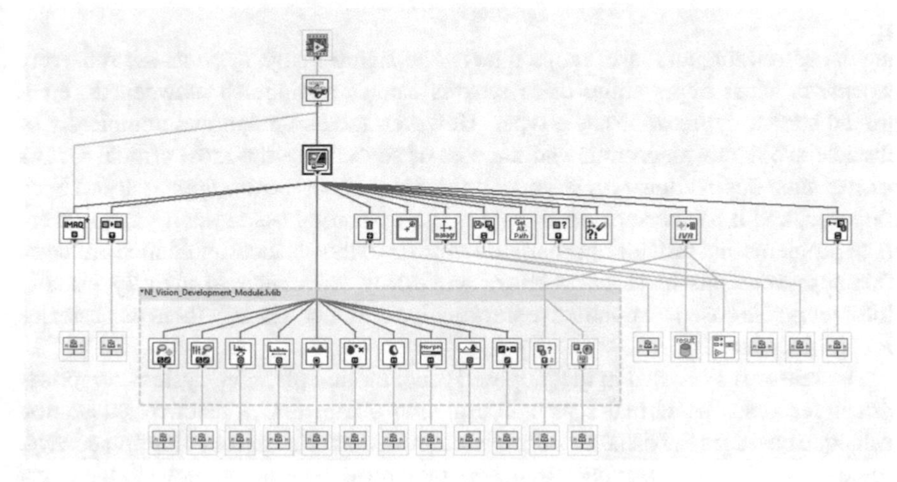

Fig. 4.14 The hierarchical representation of vision assistant express VI

source image into other grayscale values in the transformed image details in areas containing significant information at the expense of other areas. With this method, the gray level values of the pixels are changed so that the defined grayscale is evenly distributed in the range 0 to 255 for an 8-bit image to increase the contrast in the images. The 'Inverse' function reverses the image obtained in the previous step. All the steps performed on the image is shown in Fig. 4.13.

The hierarchical structure of the functions used in the image processing application examined in the titles described so far is shown in Fig. 4.14.

4.1.3.5 Vision-Based Obstacle Avoidance

Obstacle avoidance and robot navigation are essential research problems. For this problem, ultrasonic sensors, laser distance meters, and stereo vision techniques are sensing hardware components used in distance-based obstacle detection. All of these have advantages and disadvantages. For example, ultrasonic sensors suffer from low

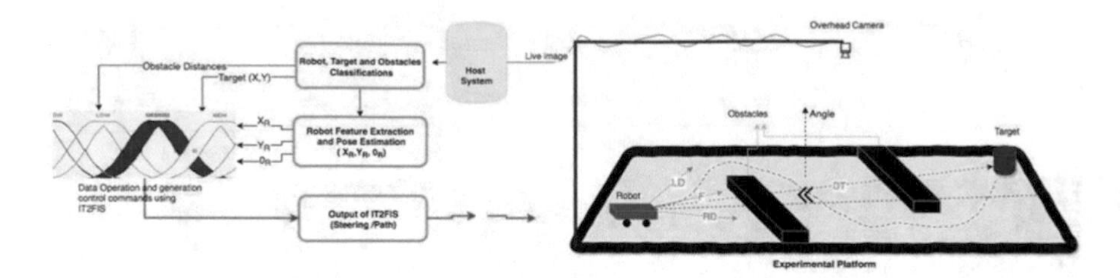

Fig. 4.15 The system architecture of visual servoing-based path planning for a mobile robot (DT: distance from the target, Distance from obstacles (Left Distance (LD), Front (F) and Right Distance (RD), ---: robot path)

angular resolution, and laser range finders and stereo image systems are relatively expensive. Most importantly, these sensors cannot distinguish between different ground surface types or obstacle types. However, the computational complexity of obstacle avoidance algorithms and the cost of sensors are the most critical factors for real-time applications. Regarding all these situations, a new approach has been proposed, and it has been decided that the use of vision-based systems can prevent these problems and provide appropriate solutions to the obstacle avoidance problem. This approach consists of basic image processing techniques to identify visually different pixels based on qualitative information and then classify them as obstacles or non-obstacles [16].

The sensor fusion used in the proposed robot motion planning system comprises virtual sensors. The virtual sensor information obtained is a reactive calculation technique based on the distance information from obstacles in robot motion control. These sensor data are the input parameters that cause the robot to move to the target and avoid obstacles.

The proposed visual-servoing (VS) based mobile robot global path planning is realized using virtual inputs generated entirely from image information. The obtained visual information enables us to generate the robot path. The graphical representation of the proposed technique is illustrated in Fig. 4.15.

In the VS control loop, the virtual sensors are the measured distance between the robot and the obstacles surrounding the robot. These are the distances to the left (LD), right (RD), and front (FD) obstacles. These sensor values and robot position information (xR, yR, θR) have been served as the inputs to the controllers to enable the generate the desired obstacle-free path. The steps of image processing application and sub-steps of morphological processes used in path planning applications are shown in Fig. 4.16. Explanation of the image processing functions used in this figure is given in the previous chapters.

Several formulas and mathematical calculations are required to obtain the input parameters of path planning algorithms. After determining the starting position of the mobile robot and the target point, other calculations are performed. The required parameters are characterized in Fig. 4.17. The formulation of these parameters is given below.

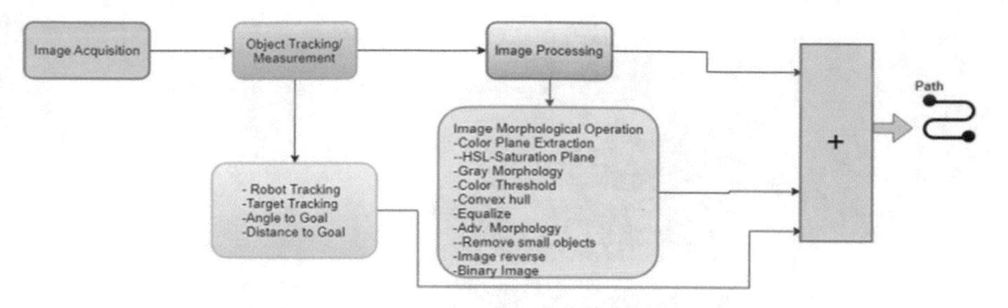

Fig. 4.16 Structure and functionality of vision-based path planning algorithm

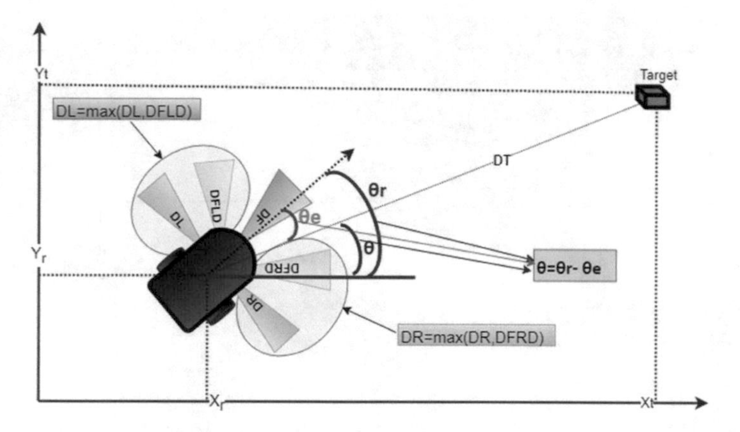

Fig. 4.17 Tracking of a virtual reference path (angle to the goal (θ = θr- θe), distance from the target (DT), distance from the obstacle (RD: right distance, F: front distance, LD: left distance, DFRD: distance from right diagonal, DFLD: distance from left diagonal)

Assuming that the mobile robot's initial position information and target position are known, as shown in this figure. In this case, the positioning error calculations are calculated as in Eq. (4.12).

$$\begin{aligned} e_x &= X_t - X_r = DT * \cos(\theta\mathrm{r}) \\ e_y &= Y_t - Y_r = DT * \sin(\theta\mathrm{r}) \end{aligned} \tag{4.12}$$

where, DT corresponds to the current distance between the mobile robot and target, which is expressed in Eq. (4.13).

$$DT = \sqrt{(e_x)^2 + (e_y)^2} \tag{4.13}$$

The robot current angle (θr) according to the target is computed as in Eq. (4.14).

$$\theta\mathrm{r} = tan^{-1}\frac{e_y}{e_x} \tag{4.14}$$

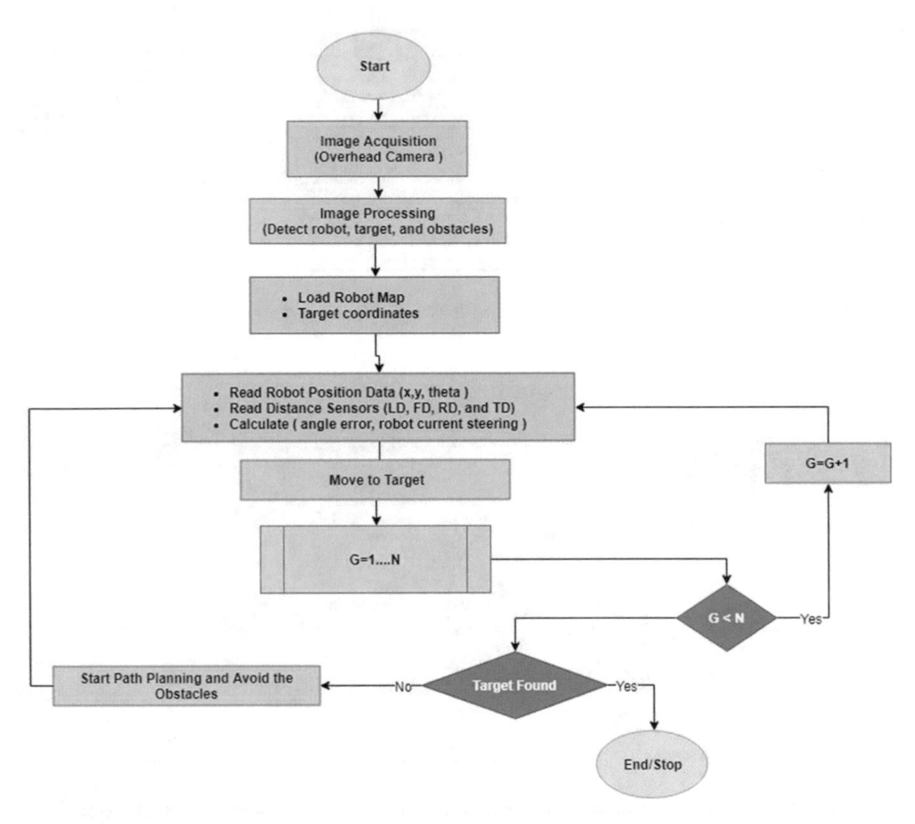

Fig. 4.18 Block diagram of path planning and obstacle avoidance system

The error of the angle is given in Eq. (4.15).

$$\theta e = \theta r - \theta \tag{4.15}$$

The flowchart of the mobile robot path-planning algorithm and the block diagram of the system integration are shown in Fig. 4.18. The abbreviations given in this block diagram have the following meaning. G shows the processed image frame, DT, RD, FD, and LD are the target, right, front, and left distances, respectively.

4.2 Vision-Based Obstacle Free Path Planning Algorithms

In the previous section, the information on the obstacle-free navigation system requirements is provided. In this section, path planning strategies and path planning algorithms will be discussed. This strategy is indicated by number 2 on the developed LabVIEW front panel (see Fig. 4.9). The image obtained in the first stage is processed at this stage, and the location, size of the working environment map, and information about other environment objects (robot, obstacles, and target) are acquired. In image processing, the functions of multiple stages were used. First, by performing the color template matching process; the initial position information (x, y, θ) and target coordinate of the robot were determined. Then a color-based filter was applied, and obstacles were identified. Here is the benefit of using template matching: In our previous work, we have also worked [17] color-based approaches. However, color-based approaches have disadvantages. If the robot and other objects were of similar color or matched at a different light intensity, errors could also occur. The color template matching method was used to overcome this problem. Thus, it is aimed to detect the labels placed on the robot and to minimize the possible errors.

After determining the center coordinates of the objects (Right, Left, Front, and Target labels), a color filter is applied to determine the positions of the obstacles. Here, the color RGB image is converted to HSL color space, and a color filter is applied. Then, the necessary input parameters for path planning algorithms were obtained. The structure of the obstacles obtained in the working area (such as sharp edge, center-blank) is valid on the success of the applied algorithm. The convex hull method has been applied to prevent problems such as local minimum issues or collisions with sharp edges. The details of the implementation steps outlined so far for each title are given above.

The vision-based obstacle-free path planning and navigation strategy have been classified based on the prior information of the environment required for path planning [18]. It is broadly classified into two categories; (i) local path planning, (ii) global path planning. In global path planning, the prior information of the environment, obstacle position, and goal position have completely known, thereby robots can reach the target by following a predefined path. On the other hand, in local path planning, there is no need for prior knowledge of the environment, and the robot has no or partial knowledge of the navigational environment [19]. The necessary steps involved in the path planning scheme is given in Fig. 4.19.

Many algorithms make use of Autonomous Mobile Robots (AMR) to be smart tools for reaching their target position in an optimized way without bumping obstacles based on criteria such as distance, time, or energy. The less computational complexity and less power consumption are essential criteria in the robot navigation system. There are many algorithms proposed in the literature for global or local path planning that are listed in Table 4.1. They are divided into two groups; (i) the classical approach, (ii) local navigation approaches, or reactive approaches, which are more intelligent and can autonomously control and implement a plan [19]. In this work, we have highlighted and used both classical and reactive approaches based on their effectiveness and application for the specific environment. After giving brief information about the algorithms we use, the results of their application, and the comparison of their effects on the selected experimental configurations are given in the following sections.

We have conducted many experimental studies on path planning. In this section, the results of five of these studies are presented. These experimental configurations are designed in different robot direction and obstacle positions. These five

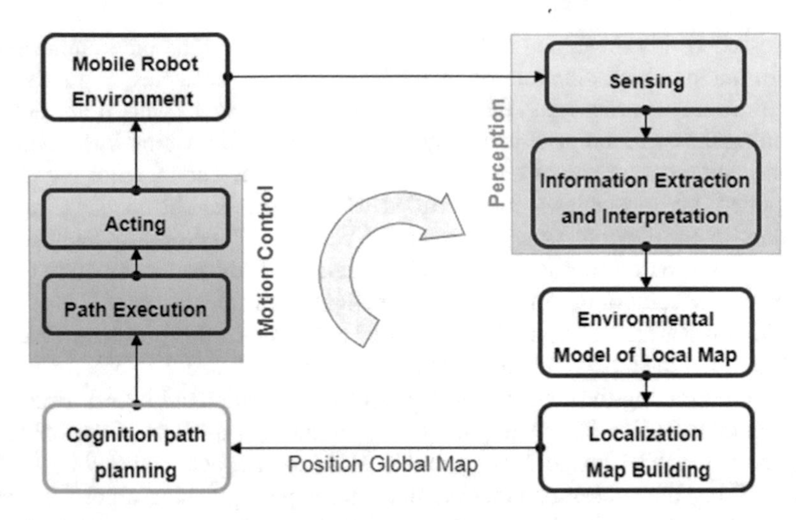

Fig. 4.19 A block scheme of the navigation system [20]

Table 4.1 Classification of different autonomous mobile robot (AMR) navigational algorithms

Reactive algorithms	Classical algorithms
Fuzzy Logic	Artificial Potential Field
Genetic Algorithm	RRT (Rapidly Exploring Random Trees)
Neural Network	B-RRT (Bidirectional RRT)
Particle Swarm Optimization	PRM (Probabilistic Roadmap)
Ant Colony Optimization	VFH (Vector Field Histogram)
Cuckoo Search Algorithm	A* algorithm
Artificial Bee Colony Algorithm	Roadmap Cell Decomposition Approach
Other Miscellaneous algorithms	Grid-Based Methods

experimental environments and processed images or robot maps are shown in Fig. 4.20—a configuration of Experimental environments. Information about the image processing steps performed here is not provided because this process was discussed in previous chapters. The following planning algorithms were implemented on these experimental configurations, and the results obtained were discussed.

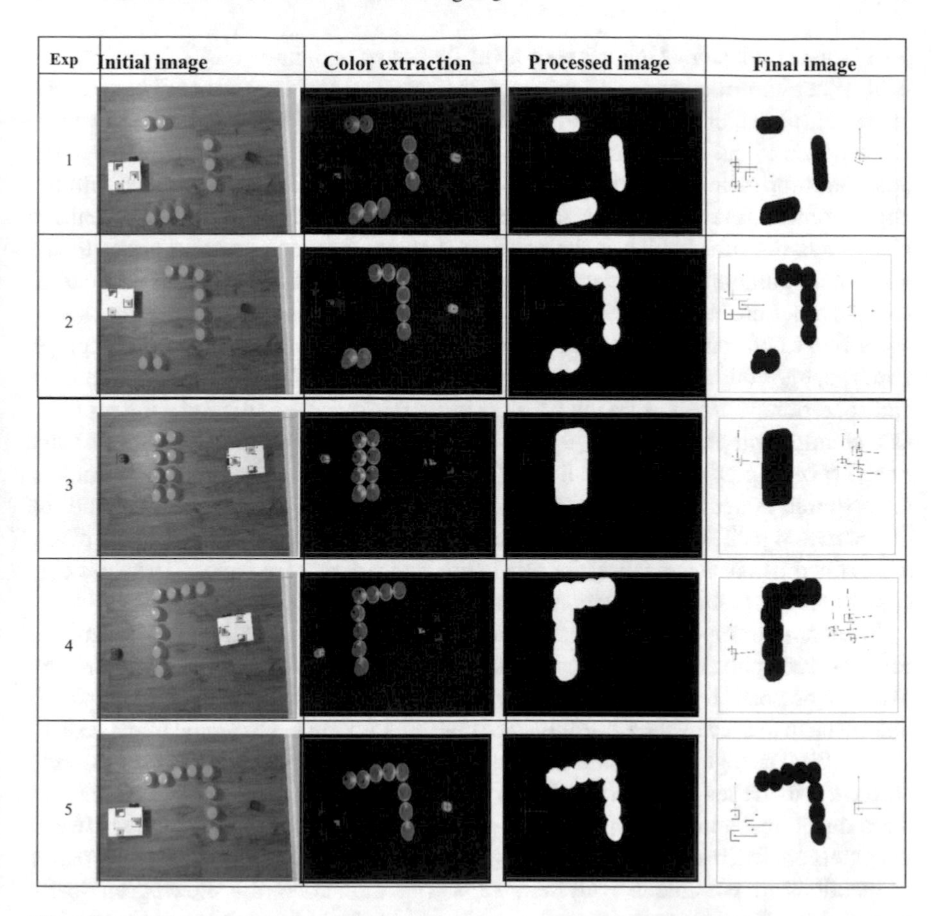

Fig. 4.20 Configuration of experimental environments

4.2.1 A Algorithm*

A star algorithm proposed by Haret et al. [21]. The A* algorithm is a practical search algorithm for path-finding and graph traversals in the real-world problem, which is a class of intelligent search algorithms in the Uniform Cost Research (UCS) philosophy developed based on Dijkstra [22] algorithm that it can find the shortest path. The key of the A* algorithm is to establish the evaluation function given in (4.16).

$$f(n) = g(n) + h(n) \tag{4.16}$$

where f(n) represents the expected cost f(n) from source to goal via node n, g(n) represents the exact cost of the path from the starting point to any vertex n, and h(n) represents the heuristic estimated cost from vertex n to the goal. The specific domain information in the problem is the heuristic function, which is an estimated distance of the node n to the goal. The Euclidean distance (ED) between the node n and the goal is usually taken as the value of h (n) that is an estimated cost of reaching the goal. When the value of g(n) is constant, the value of f(n) is mainly affected by the value of h(n), which is the cost value from the successor node to the destination node corresponds to the Manhattan distance (heuristic). The A* algorithm considers the position information of the mobile robot's target point and searches along with the target point. For an application such as routing, h (n) can physically represent the bird' s-eye distance, which is the smallest distance between any two nodes to the target configuration. In the algorithm, the search is shaped according to the state of the cost function. Nodes with a low-cost function are preferred over larger nodes, and according to this philosophy, routing continues. The algorithm is optimal as a graph search using both an open and a closed set of nodes while ensuring acceptability and consistency. When using the A* algorithm, it is necessary to model the problem as a standard graphical search algorithm. In our application, the converted binary image (pixel graph) is used as an input parameter for the algorithm. All regions of the acquired image pixels are searched one by one to find the shortest path, and the unobstructed path from the source to the destination is determined. All black pixels are defined as obstacles; all white pixels are defined as a free node. The total cost constitutes the evaluation function's cost calculated between the free nodes.

The size of the open-and-closed list can cause the efficiency of this algorithm to fall for real-time applications and is useful during the computation time of the algorithm. That can be normalized by changing the received image resolution or pixel graph. To obtain the map used by the algorithm, the image resolution in our applications is set to 100×100. The higher the resolution of the map, the better the results, but undesirable because it increases computational time in real-time applications [23]. Each pixel of the reduced resolution map is taken as a corner, and the connection paths between the pixels are taken as the edge. For any general position of the robot, possible matrix connections are possible in 3 different ways. These situations are shown graphically in Fig. 4.21. Connection matrices ((a) only allows the robot to move linearly (up, down, left, and right), (b) permits the robot to take four transverse motions with four linear motions, (c) also allows the robot to perform more flexible movements by inserting connections between transverse movements). All possible movements are indicated by 1, and impossible movements by 0. For increasing the rotational flexibility of the robot, the Cardinality of numbers in the matrix can be increased,

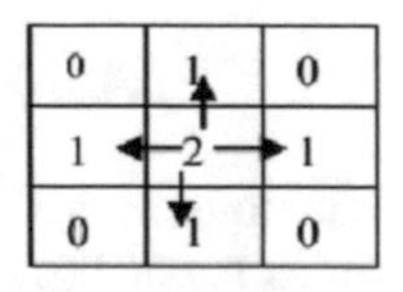

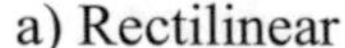

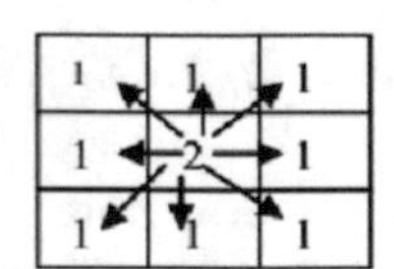

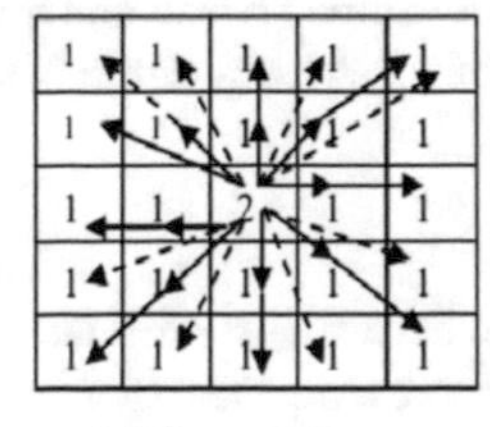

Fig. 4.21 Connection matrices (**a** only allows the robot to move linearly (up, down, left, and right), **b** permits the robot to take four transverse motions with four linear motions, **c** also allows the robot to perform more flexible movements by inserting connections between transverse movements)

but the addition of these can result in more calculation costs. In our application, the "Rectilinear and Diagonal" matrix was used.

The experimental setup takes the map as the filled grid of the binary image. The experimental results of the occupancy grid image of 1024×576 rectangular pixels are shown in Fig. 4.20.

As a binary image which is a 2D matrix of elements that can only hold two values where the white pixels (values 1) correspond to the free space and the black pixels (values 0) correspond to an obstacle area, it suffices to represent a fill grid with only two color levels, since the robot can only move within the free space. The obstacle-free path and other operations obtained by using this algorithm are shown on eight experimental results (see Fig. 4.22). These experiments were determined by using the experimental configurations in Fig. 4.20.

To thoroughly test the behavior of the algorithm, several maps were used; only the results of the eight maps are shown here. The processing time and path length calculated here are obtained from the 1024×576 size maps with various obstacles. Figure 4.20 shows the optimum path between the start and goal obtained by using A* path searching algorithm, where the searched space or pixels are also marked. To obtain a feasible path that the robot can follow, it is aimed to reduce the possibility of collision during the movement of the robot by using dilatation and convex hull operation.

For this reason, the obstacle boundaries have been extended to the radius (half dimension) of the actual robot and the robot's safe operation without collision has been realized. In this case, the environment map has been changed so that the free space near the obstacles can be regarded as a disability area at distances below the radius. In the test cases, it is observed that the algorithm was able to find a feasible path solution to use by any robotic controller to move the robot physically.

4.2.2 RRT

A rapidly exploring random tree (RRT) is a search algorithm with a single-query tree structure based on uniform random sampling, which is the start and goal point are certain known. The RRT algorithm grows based on the configurations of construction, a tree seeking the target point from a starting point, where each node of the tree is a point (state) in the workspace [24–26]. This algorithm was introduced by LaValle [27]. The RRT method first starts a roadmap configuration with a starting point (q_{start}) as a tree root [28, 29]. Next, the algorithm selects a random point (q_{rand}) for each next iteration in the configuration space, and the nearest node (q_{near}) from the existing graph is searched. A new sample (q_{new}) is generated with a predefined distance (ε), namely a distance of step size (step size = 20 px), from q_{near} to q_{rand}. A collision is considered in the newly selected node (q_{new}). If a collision occurs, a new step (q_{new}) is discarded. Otherwise, it is added to the search tree so that for each new point, the distance between the newly generated node and the target point is considered [30]. The procedure is illustrated in Fig. 4.23.

Exp.	Initial Image	Expansion	Path	Path Length/ Exec. Time (sc)
1				ET=5.002872e+00 PL=7.569954e+00
2A			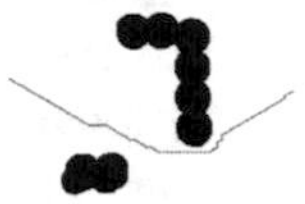	ET=3.912832e+00 PL=8.005619e+02
2B				Convex Hull appl. ET=4.505114e+00 PL=7.932931e+02
3				ET=8.046721e+00 PL=7.620224e+02
4A				ET=3.984761e+00 PL=7.569061e+02
4B				Convex Hull appl. ET=3.416262e+00 PL=7.368838e+02
5A				ET=5.194934e+00 PL=7.702028e+02
5B				Convex Hull appl. ET=3.279009e+00 PL=7.702028e+02

Fig. 4.22 Experimental sample results using A* algorithm (A in (2, 4, 5) is the normal environment; B in (2, 4, 5) is convex hull applied environment; ET: Execution Time; PL: Path Length)

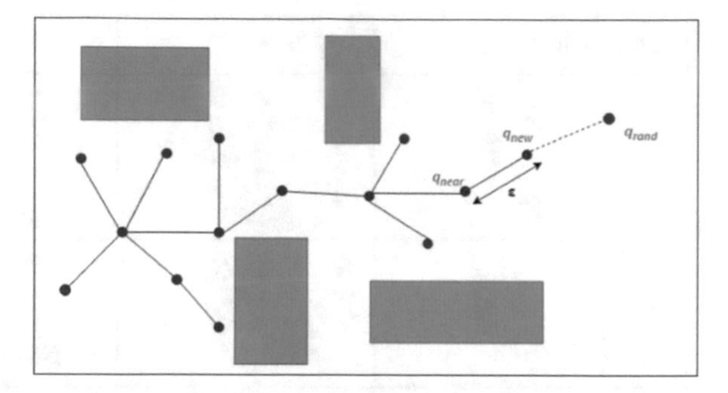

Fig. 4.23 Extension of the RRT graph using the straight-line local planner with a resolution (ε)

When searching in the larger configuration area, the RRT is very time-consuming due to its random growth. It is crucial to create a purposeful search tree. That can be accomplished by selecting the goal as the sample for bias percentage of iterations, rather than randomly setting up the sample [24]. The search for the growth of the tree will continue according to the predetermined threshold. The path is found if the distance is less than or equal to the set threshold. Otherwise, the distance is repeated until the distance is less than or equal to the specified threshold value, or the call times exceed the number of iterations. It can be considered using the forward tree expansion motion model, which allows the RRT to find viable trajectories even under differential constraints. That is one of the advantages and reasons for the extensive use of RRT in robotics for motion planning of systems such as mobile robots [31]. The applications that we have performed using this algorithm and their results are shown in Fig. 4.24.

The results of these experiments indicate that the path obtained here is not suitable for the mobile robot to follow because it has sharp passes and a convoluted structure. To improve this situation or to obtain more suitable path coordinates, the Dijkstra algorithm was applied to the results obtained from the RRT algorithm. Thus, the cost of the path has been reduced and the condition of the path has been made more feasible for path tracking. The Dijkstra algorithm is an algorithm that finds the shortest paths from a source node to all other nodes in the graph, separated from the A* algorithm by not using the heuristic function [25].

The disadvantage of this algorithm is that it takes into account all nodes between the beginning and the target node and is not intuitive. The result of the use of Dijkstra's algorithm over experiments is given in the same environment with the RRT algorithm together. It is called after the RRT operation before path tracking starts. Here, it is used as an alternative path planning to eliminate the event of roundabout presence paths and sharp passes. It is clear that due to the use of Dijkstra and RRT together, the total length of the path decreases while the time spent in path planning is prolonged.

Exp.	Initial Image	RRT Tree	Path	Path Length/ Exec. Time (sec)
1		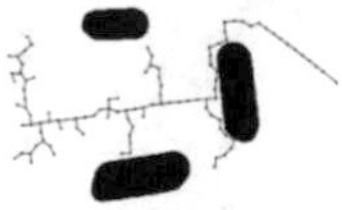	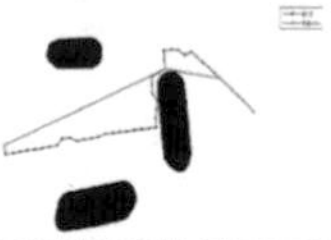	RRT path ET=1.144,635e+01 PL=8.292435e+02 RRT+Dijkstra Path ET=1.595473e+01 PL=7.367822e+02
2A		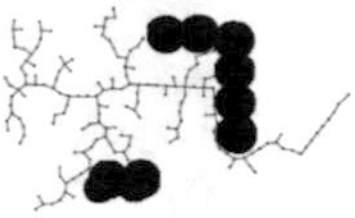	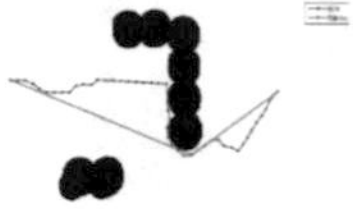	RRT path ET=1.395314e+01 PL=8.986829e+02 RRT+Dijkstra Path ET=1.753977e+01 PL=7.680722e+02
2B		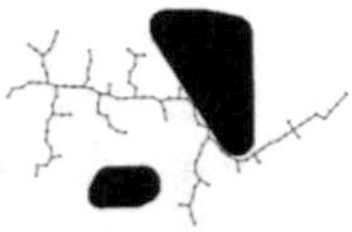		RRT path ET=1.104716e+01 PL=8.999475e+02 RRT+Dijkstra Path ET=1.320736e+01 PL=7.679939e+02
3		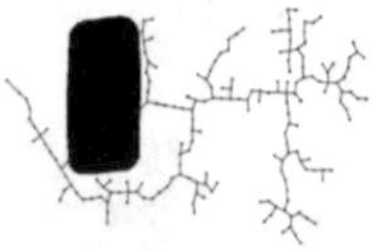	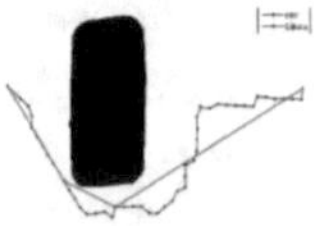	RRT path ET=2.546621e+01 PL=1.048694e+03 RRT+Dijkstra Path ET=2.938200e+01 PL=8.012886e+02
4		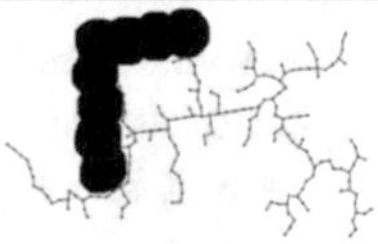	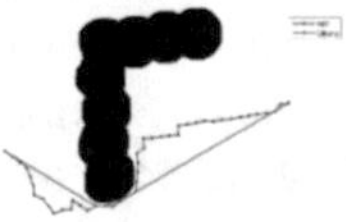	RRT Path ET=2.289973e+01 PL=8.937618e+02 RRT+Dijkstra ET=2.720643e+01 PL=7.303033e+02
5A			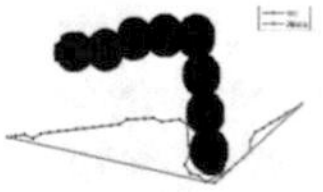	RRT Path ET=1.603056e+01 PL=9.021969e+02 RRT+Dijkstra ET=1.766586e+01 PL=7.416814e+02
5B				RRT Path ET=1.866617e+01 PL=9.124474e+02 RRT+Dijkstra ET=2.058875e+01 PL=7.350454e+02

Fig. 4.24 Experimental sample results using RRT and Dijkstra algorithms (2A, 4A, 5A is a normal environment; 2B, 4B, 5B is convex hull applied environment; ET: Execution Time; PL: Path Length)

4.2.3 BRRT

The BRRT (*Bidirectional-RRT*) algorithm is an algorithm that works in the form of growth and development of two trees towards each other at the source and destination node, rather than just one tree [32]. It is an advanced version of RRT, which also explores the search space using trees, starts branching from source and target position at the same time, and searching for each-other [24]. This assumes that the reverse path from the target to the source can be calculated, assuming that the actions are reversible or that the plan is intentionally knowing from source to destination. The expansion

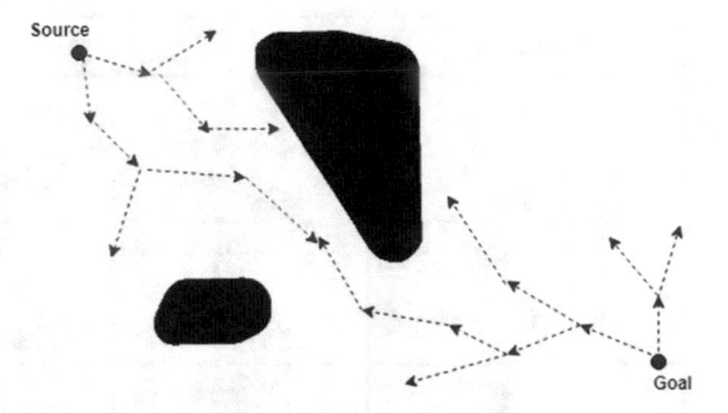

Fig. 4.25 Bi-directional RRT (BRRT) algorithm

and growth of both trees will grow toward each other and expand by random steps and a number of factors (bias value). In the end, the algorithm terminates when both the trees come together. The concept is shown in Fig. 4.25.

Based on this philosophy, we carried out experimental studies. Dijkstra algorithm was used in these studies as in the RRT. The aim is to make the obtained path information more usable and to allow the robot to follow the path with minimum possible errors while performing real-time path tracking applications. The results of the experiments are shown in Fig. 4.26. This figure shows both the path length and processing time of BRRT and the path length and processing time of Dijkstra + B-RRT.

4.2.4 PRM

The PRM (probabilistic roadmap) algorithm is used for path searching between the randomly selected points, which is in the workspaces as the vertices are distributed samples stored in the path map are collision-free samples. It is a free space construction for a path map or undirected graph. The randomly selected points (vertices) need to be out of the obstacles. Between the source and goal configuration points, some graph search algorithm like A* and Dijkstra is used to finding the optimum path with making the connection, which is connected to their neighbors by a straight is known as an edge, of all vertices [30]. The specific visibility points (collision-free points) are added to the map to construct the path. The path map is initially an empty cluster. That is the representation of the configuration space Qn of a robot with a graph G (V, E). The vertices of the graph $V \subset Q$ represent the set of configurations that are used by the planner as waypoints [33]. A randomly chosen sample configuration q_{rand} in Qn is included in the roadmap vertices (V), and nodes in the workspace that are needed to expand the map are found. All nodes can be found by choosing the K-nearest neighbor or whose distance is smaller than the pre-defined q_{rand} parameter (D). The edges (E) of neighboring vertices connect if the edges are collision-free. Depending on the size of the map, the number of random nodes will connect with neighboring nodes so that the final path can be determined. If a path cannot be found

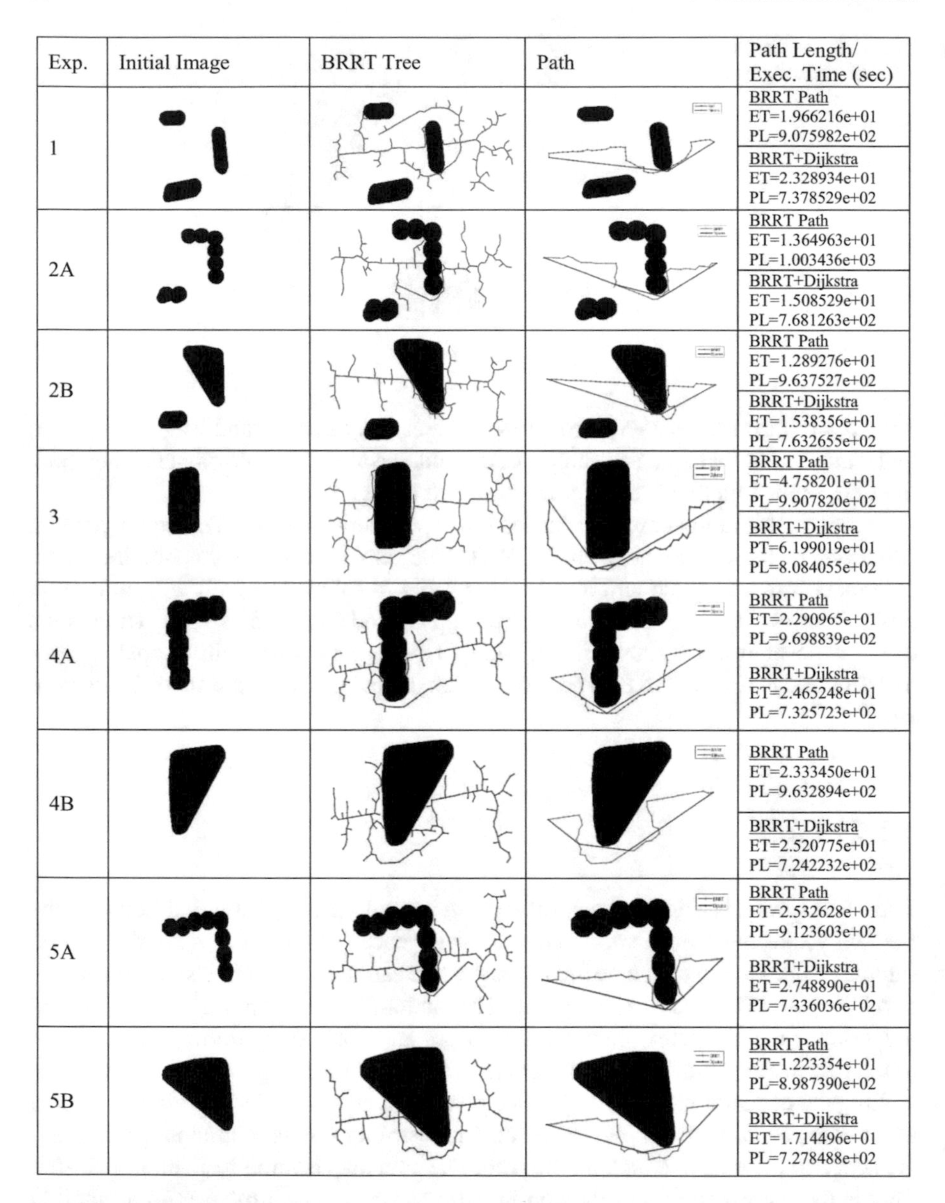

Exp.	Initial Image	BRRT Tree	Path	Path Length/ Exec. Time (sec)
1				BRRT Path ET=1.966216e+01 PL=9.075982e+02
				BRRT+Dijkstra ET=2.328934e+01 PL=7.378529e+02
2A				BRRT Path ET=1.364963e+01 PL=1.003436e+03
				BRRT+Dijkstra ET=1.508529e+01 PL=7.681263e+02
2B				BRRT Path ET=1.289276e+01 PL=9.637527e+02
				BRRT+Dijkstra ET=1.538356e+01 PL=7.632655e+02
3				BRRT Path ET=4.758201e+01 PL=9.907820e+02
				BRRT+Dijkstra PT=6.199019e+01 PL=8.084055e+02
4A				BRRT Path ET=2.290965e+01 PL=9.698839e+02
				BRRT+Dijkstra ET=2.465248e+01 PL=7.325723e+02
4B				BRRT Path ET=2.333450e+01 PL=9.632894e+02
				BRRT+Dijkstra ET=2.520775e+01 PL=7.242232e+02
5A				BRRT Path ET=2.532628e+01 PL=9.123603e+02
				BRRT+Dijkstra ET=2.748890e+01 PL=7.336036e+02
5B				BRRT Path ET=1.223354e+01 PL=8.987390e+02
				BRRT+Dijkstra ET=1.714496e+01 PL=7.278488e+02

Fig. 4.26 Experimental sample results using BRRT and Dijkstra algorithms (2A, 4A, and 5A is the normal environment; 2B, 4B, and 5B is convex hull applied environment; ET: Execution Time; PL: Path Length)

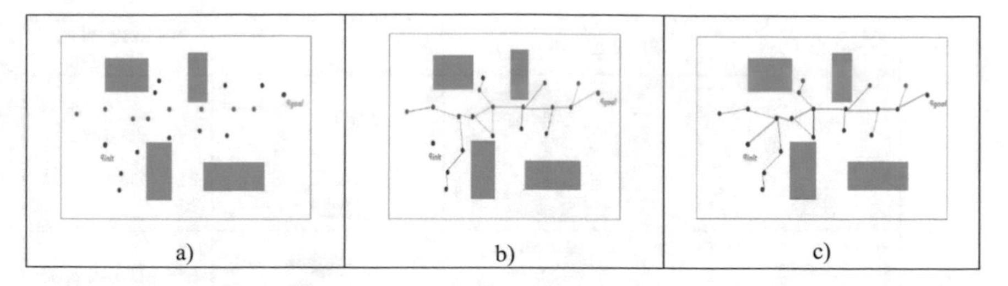

Fig. 4.27 The configurations sampled in the first phase of the PRM (red). Initial and goal configurations are not necessarily sampled (**a**). Roadmap constructed from the samples using the straight-line planner (**b**). In the query phase, both start and goal configurations are connected to the roadmap and a path is found (**c**)

in the first iteration, this parameter (sample nodes) must be changed and retried. As the number of nodes increases, it may be possible to find the optimal path, but the processing time increases. We can change any number of nodes until the map finds the desired path. The example of a roadmap construction is depicted in Fig. 4.27.

In the appropriate route-finding stage, the A* search algorithm was used to find the best possible path between the desired start and the target node. In our experiments, taking into account the size of the map, the number of nodes is determined to be 50. Although the number of nodes is related to the working area, the configuration of the obstacles in the environment is also valid [33]. The results of our experiments are shown in Fig. 4.28. Both the path length and the application time were calculated here.

The second column of Fig. 4.26 shows the initial image with vertices. To determine the appropriate number of vertices for the algorithm, it may be necessary to run the algorithm several times by changing the number of samples (number of nodes) in the working space. If the algorithm has a longer working time, the number of points/vertices can be reduced. If the algorithm gives results instantly, the number of nodes may wish to increase these samples to hope to get a better path.

4.2.5 APF

In this section, we analyze traditional APF approaches and explain in more detail why we recommend this method with VBC (vision-based control). In the field of artificial potential field, robot motion is shaped according to two parameters: (i) attractive force and (ii) repulsive force. These attractive and repulsive forces are vector magnitudes that are consists of magnitude (m) and direction (d) parameters. Its main philosophy is that the same polar charges push each other in the electromagnetic field, and the opposite polar loads attract each other, as shown in Fig. 4.29a. It solves the problem by assuming that all obstacles are the source of repulsive potential, assuming that the potential is inversely proportional to the distance of the robot from the obstacle.

Exp.	Initial Image	PRM (Road map)	Path	Path Length/ Exec. Time (sec)
1	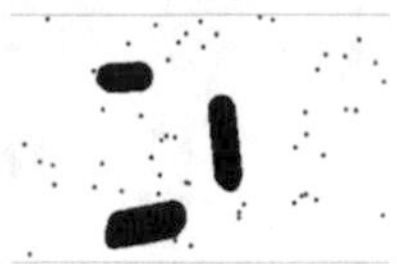	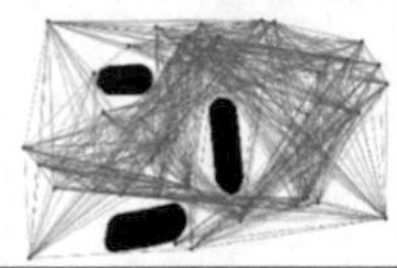		ET=3.832857e+01 PL=7.778271e+02
2A	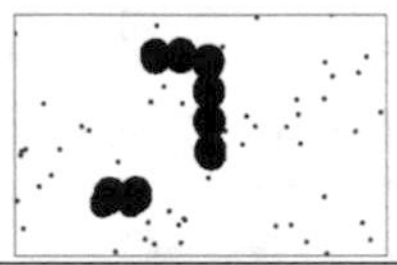		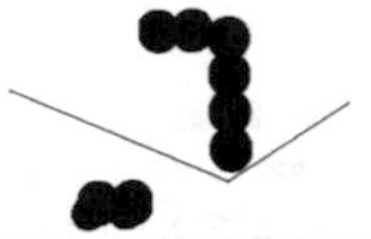	ET=4.194708e+00 PL=8.432752e+02
2B		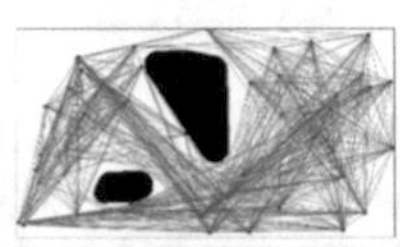		ET=2.810440e+00 PL=8.946148e+02
3	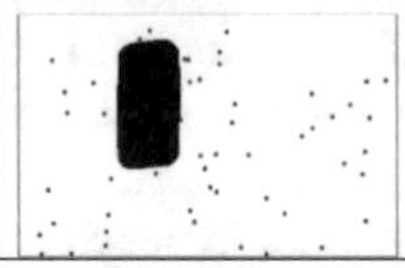			ET=4.193119e+01 PL=8.162892e+02
4	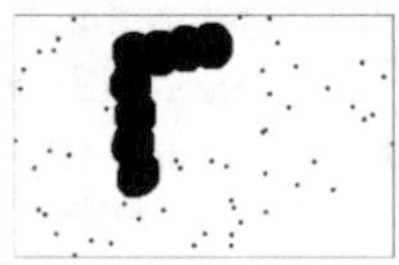			ET=3.910240e+00 PL=7.402907e+02
5A	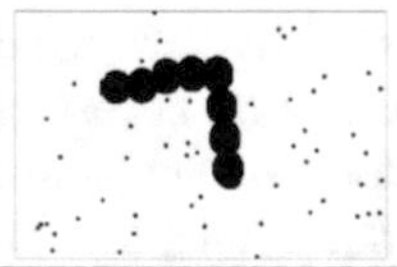	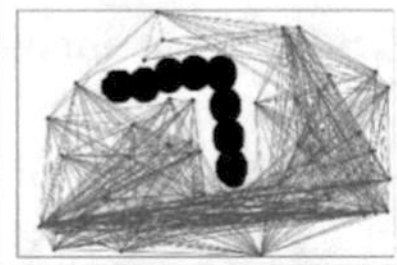		ET=5.298011e+00 PL=7.538851e+02
5B	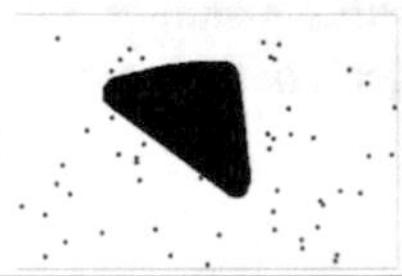	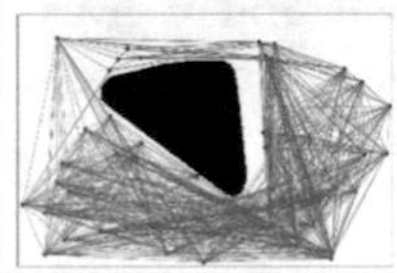	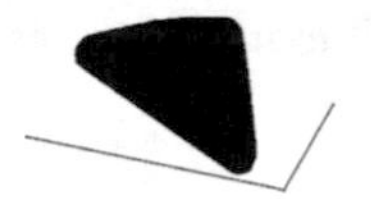	ET=6.123063e+00 PL=7.996370e+02

Fig. 4.28 Experimental sample results using PRM algorithm (A in (2, 5) is the normal environment; B in (2, 5) is convex hull applied environment; ET: Execution Time; PL: Path Length)

The target attracts the robot by applying an attractive potential. The derivative of the potential gives the value of the virtual power applied to the robot, based on where it moves. The magnitude of this force can generally show variability according to the size of the obstacles and distance values between objects and obstacles. It is important to note that the robot and the target are loaded with the same polar load. In a potential field, a robot can be treated as a point in a 2D environment [34].

A single purpose of attractive potential is to attract the robot to the target. The vector F1 can represent this gravitational force in Fig. 4.29. Various parameters can be

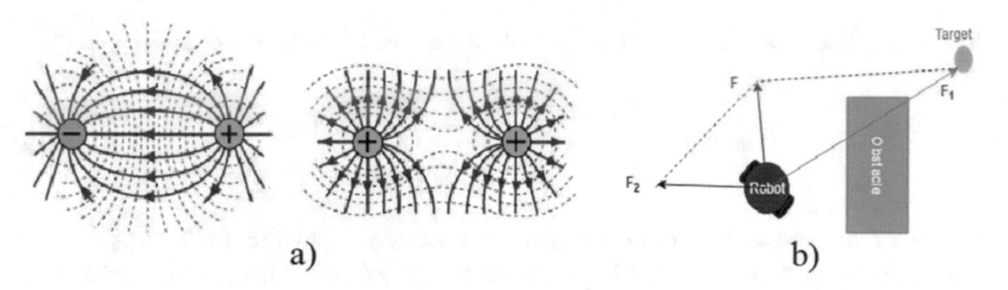

Fig. 4.29 Definition of attractive force and repulsive force in an artificial potential field

used to model this situation. The most critical parameter is the distance between the robot's current position and the target position. The degree of proportionality and the proportionality constant are also the parameters used. These are algorithm parameters that can be set for different purposes, such as controlled adjustment of the speed of the robot as it approaches the target, as well as maintaining high clearance and short path lengths. In a configuration space (2D), the robot is represented as a particle under the influence of an artificial potential field P, whose variations reflect the structure of the free space P (p_r) for a given configuration $p_r = (x, y)^T$ where p_r is representing the robot position. The repulsion potential is applied by all obstacles that prevent the robot from approaching them and potentially cause a collision. The sum of an attraction potential $P_{att}(r)$ is pulling the robot towards the goal configuration, and a repulsion potential $P_{rep}(r)$ is pushing the robot away from obstacles [35]. The sum of these two functions can be defined in Eq. (4.16).

$$P_{total}(p_r) = P_{att}(p_r) + P_{rep}(p_r) \tag{4.16}$$

In this equation, the attractive potential field function is formulated in Eq. (4.17).

$$P_{att}(p_r) = \frac{1}{2}k_a\left(p_r - q_g\right)^2 \tag{4.17}$$

where p_r is the coordinate of the robot, k_a is the coefficient constant of the attractive field, and q_g is the goal coordination indicator.

The robot (particle) in the space moves towards the target point under the gravity generated by the target point. Therefore, the gravitational force on the particle can be expressed as the negative gradient equation of the target point potential field [35]. The gradient artificial force vector field F (q) is defined as Eq. (4.18).

$$F(p_r) = -\nabla P_{att}(p_r) + -\nabla P_{rep}(p_r) = F_{att}(q) + F_{rep}(q) \tag{4.18}$$

In this equation, ∇P expresses P gradient vector. $F_{att}(p_r)$ represent the attractive artificial force and $F_{rep}(p_r)$ is the artificial repulsive force.

In artificial potential, obstacles create repulsive areas. When the robot is far enough away from obstacles, it does not consider the propulsion force for movement towards the target. The repulsion potential field function can be formulated as Eq. (4.19).

$$P_{rep}(p_r) = \begin{cases} \frac{1}{2}k_r\left(\frac{1}{d(p_r)} - \frac{1}{d_{max}}\right)^2, \ d(p_r) \leq d_{max} \\ 0, \ d(p_r) > d_{max} \end{cases} \tag{4.19}$$

where k_r represent the coefficient constant of the repulsion field, d_{max} is the maximum impact extent of the single obstacle, $d(p_r)$ is indicate the distance between the robot and obstacle.

The repulsive force function is the negative gradient of repulsive function as formulated in Eq. (4.20).

$$F_{rep}(p_r) = -\nabla P_{rep}(p_r) \begin{cases} \frac{1}{2}k_r\left(\frac{1}{d(p_r)} - \frac{1}{d_{max}}\right)\frac{1}{(d(p_r))^2} \cdot \frac{\partial d(p_r)}{\partial p_r}, \ d(p_r) \leq d_{max} \\ 0, \quad d(p_r) > d_{(x} \end{cases} \tag{4.20}$$

The configuration space may not always be statically or may have obstacles in different shapes and positions. It is necessary to calculate the total repulsion forces of all obstacles in the robot's range of motion. The resultant of the repulsive potential field function and force function is calculated using Eq. (4.21) and (4.23).

$$P(p_r) = P_{att}(p_r) + \sum_{i=1}^{n} P_{rep}(p_r) \tag{4.21}$$

$$F_{total} = F_{att}(p_r) + \sum_{i=1}^{n} F_{rep}(p_r) \tag{4.22}$$

where n corresponds to the number of obstacles [36].

The potential field method for robot path planning and control incorporates various combination models. While the robot movement takes place without collision, some problems such as the local minimum may be encountered. If the repulsive and attractive potential forces are equal or approximate, the robot cannot move or oscillate in any direction. This is a local minimum problem. An example of this situation is provided in Fig. 4.30. To resolve the local minima problem, there are several proposed work [37].

Using the traditional APF approach, it is necessary to develop different calculations for robotic control or path planning applications that are captured in a minimum local region, and that cannot be achieved. If the target point is too close to the obstacle, the combination of moving forces is taken into account, and the distance is gradually reduced to allow the robot to move to the target. It is also possible to develop alternative approaches based on the sum of the resultant forces (pushing and pulling forces) acting on the direction of the movement of the robot. To formulate these expressions, the following equations have been developed [38].

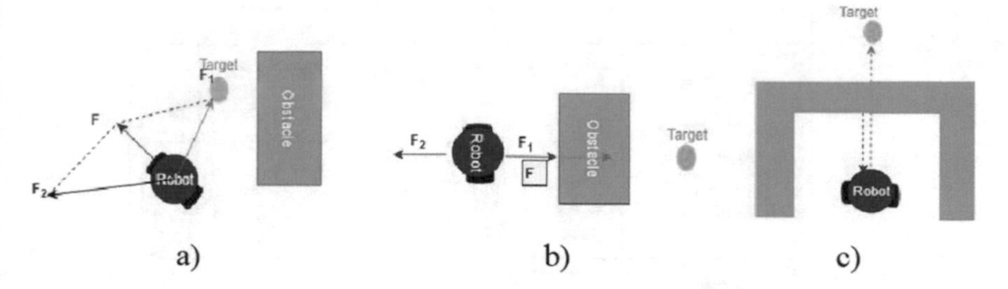

Fig. 4.30 Representative local minima in a traditional artificial potential field (APF): **a** unreachable goals near obstacle **b** trapped in a local minimum region, **c** The robot is surrounded by obstacles and the exit is opposite the goal

$$P_{rep}(p_r) = \begin{cases} \frac{1}{2}k_r\left(\frac{1}{d(p_r)} - \frac{1}{d_{max}}\right)^2 (p_r - p_t)^n, \ d(p_r) \le d_{max} \\ 0, \ d(p_r) > d_{max} \end{cases} \tag{4.23}$$

where, $(p_r - p_t)$ is the distance between the current position and the target position, k_r is the repulsion field coefficient constant, and n is an arbitrary real number. With this addition, the repulsive force function can be reduced relatively as the robot approaches the target point.

$$P_{rep}(p_r) = \begin{cases} \frac{1}{2}k_r\left(\frac{1}{d(p_r)} - \frac{1}{d_{max}}\right)^2 (p_r - p_t)^n, \ d(p_r) \le d_{max} \\ 0, \ d(p_r) > d_{max} \end{cases} \tag{4.24}$$

The resulting composite repulsive forces are calculated as in Eq. (4.25).

$$F_{rep}(p_r) = \begin{cases} F_{rep1}(p_r) + F_{rep2}(p_r) + \cdots + F_{rep(n)}(p_r), \ d(p_r) \le d_{max} \\ 0, \ d(p_r) > d_{max} \end{cases} \tag{4.25}$$

Here, $F_{rep}(p_r)$ refers to the sum of the resulting composite forces. On the axis of all these narratives, we have shown the results obtained from our experimental study in Fig. 4.31 that shown two result parameters (execution time and path length). Also, the sensor data and potential forces obtained in path planning experiments are graphically shown.

The meanings of abbreviations/legend used in Fig. 4.31: DF is Distance Front, DFLD is Distance Front Left Diagonal, DFRD is Distance Front Right Diagonal, DR is Distance Right, and DL is Distance Left, respectively. The meanings of abbreviations/legend used in the Potential Force column are the following: F_1 is a potential function of x-direction, F_2 is potential in the y-direction, and F_{rep1} and F_{rep2} are an indication of repulsive force function vector. F_{att1} and F_{att2} are indications of attraction force function vector. In our experiments, five input distance values, which are distances at specific angles are measured to compute the repulsive potential from

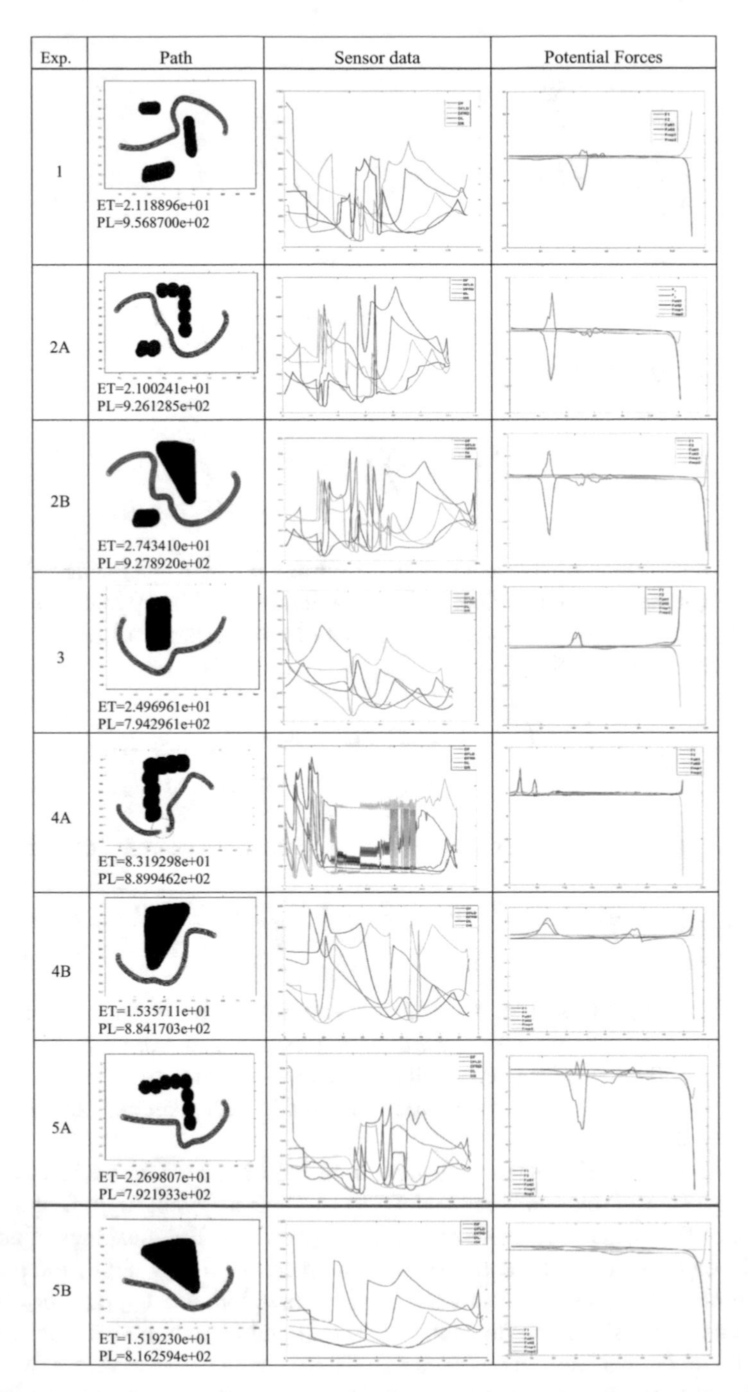

Fig. 4.31 Experimental results using APF algorithm in a static environment (2A, 4A, and 5A is the normal environment; 2B, 4B, and 5B is convex hull applied environment ET: Execution Time; PL: Path Length)

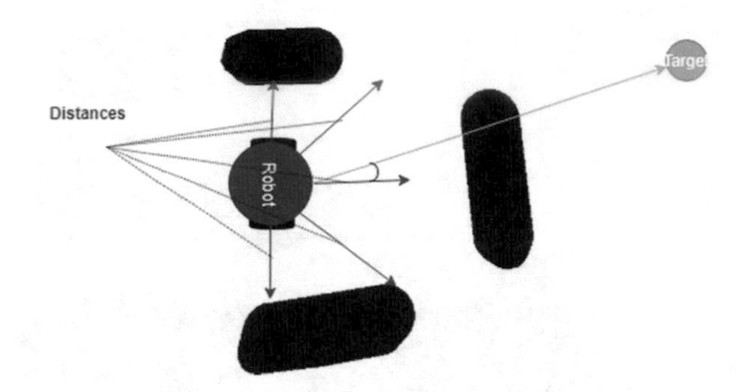

Fig. 4.32 Measurement of repulsive potential vectors

forward, left side, right side, forward left diagonal, and forward from right diagonal, and angle values according to the target is accepted as input parameters. If the distance values are less than the predetermined threshold distance value, the repulsive and attractive potential forces will be active. The combination of magnitudes on the coordinate axis of the attractive and repulsive force vectors effects the direction of motion. All obstacles repulse the robot inversely proportional to the distance, but the target attracts the robot in inverse proportion. The direction and speed of the robot take into account the potential, attractive, and repulsive components. Robot direction is indicated as the potential vector. The distance inputs are for a sample scenario is summarized in Fig. 4.32.

Several scenarios have tested in our proposed system. However, eight different test scenarios have been configured here to evaluate the performance of the system using Figure's test environments, each of which is conceived with the different levels of complexity was used. It is understood from the results of the experiments that this method is an effective method of path planning. However, the best algorithm will be determined after analyzing the obtained results together with the other algorithms we use. This evaluation is given in the experimental evaluation title.

4.2.6 GA

Genetic algorithms (GA) were first proposed by John Holland [38] as the stochastic best solution for the survival of the best in the complex multidimensional search field, similar to the evolutionary process observed in nature. GA is an effective and useful optimization method that can be used in cases where the search field is large and complex, the problem cannot be expressed in a particular mathematical model, and the desired results cannot be achieved. It is an effective way to solve real-world problems such as the mobile robot path planning based on evolutionary concepts [39–41]. GA begins with a series of solutions or populations called chromosomes.

In our experiments, the GA method has been applied by using the experimental configurations shown in the previous sections. The resolution of the robot maps in

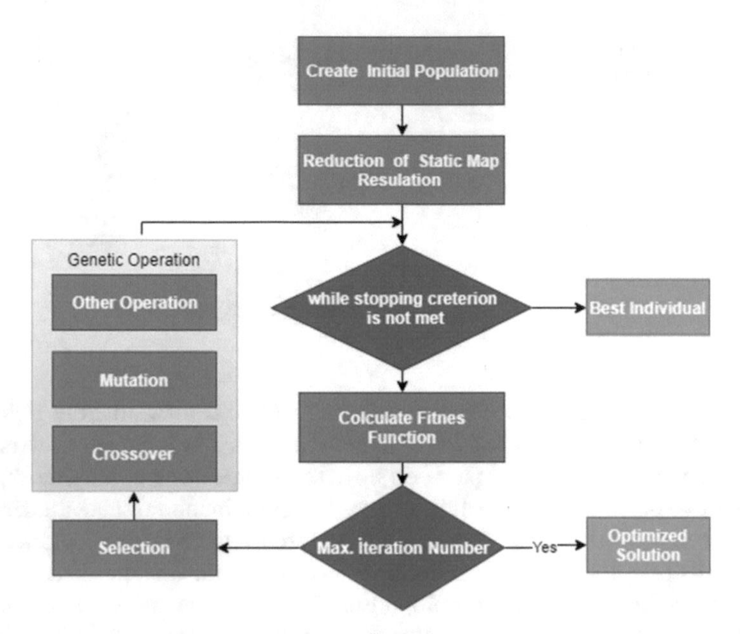

Fig. 4.33 The general flowchart of the genetic algorithm (GA)

which experimental studies are performed is high. Therefore, we first try to computationally make possible the development of the robotic path of the evolutionary algorithm by reducing the map resolution. The size of the population consists of pixels in the binary image. Each individual in the population group represents a solution assessed by the fitness function. High-grade generations are considered more appropriate than low-grade generations.

The general GA algorithm flowchart is given in Fig. 4.33. Each step of the algorithm is given in this graph (such as selection, crossover, and mutation), and other details of the algorithm are omitted. Until the desired solution is achieved, the basic algorithm follows an iterative approach to generate new generations until the specified criterion for stopping the cycle is met.

As the evolutionary search area is large, the resolution reduced to the MxN size maps by a factor. The resultant map has dimensions ceil(x/α) * ceil(y/α) that we used in the experimental configurations, therefore, the map resolution has been reduced to facilitate the calculation of the optimal path with the evolutionary approach. Each cell in the map (binary image) determined based on the presence (1) or absence (0) of an obstacle, which is aggregated value denotes the shade of gray with 1 denoting complete black and 0 denoting complete white [41]. Equations (4.26) and (4.27) are utilized to calculate these expressions.

$$C_{i,j} = \begin{cases} 0, \text{ no obstacle exists at a location(i, j) of map} \\ 1, \text{ an obstacle exists at a location(i, j) of map} \end{cases} \quad (4.26)$$

$$d_{kl} = \sum_{i} \sum_{k} C_{i,j} \tag{4.27}$$

Here $0 \leq i \leq M, 0 \leq j \leq N$ and d_{kl} is any aggregated cell of the lower resolution map. The result graphs obtained with GA, which we used to find a suitable and optimum path for the robot to reach the target from the source, are shown in Fig. 4.34.

The position of fixed points in the binary image characterized by pixels in the formation of genetic individuals is useful in determining the limit of the fitness function for optimization. The path was created by using the links between these points. The length of these points is specified for the fitness function. The graphic technique is a traditional way of representing the environment in which a mobile robot moves. In GA, the pathway is formed by midpoints of free rings represented by variable-length chromosomes. The accuracy of this method has been tested, as shown in Fig. 4.34, by creating several configuration areas.

4.2.7 Type 1 Fuzzy Logic

The fuzzy logic is one of the soft computing methods that is essentially a system to deal with uncertainty, to characterize the types of knowledge that cannot be represented by conventional Boolean algebra and it was proposed by Lotfi A. Zadeh at the beginning of the 1990s [42, 43]. The simplicity of control, low cost, and the possibility of design without knowing the exact mathematical model of the process show the importance of Fuzzy controllers. Because of these situations, fuzzy logic has become a prevalent and exciting topic of computer science and robotics and been used in a variety of applications, especially in autonomous mobile robots' navigation control applications. Mobile robot path planning in unknown (indoor/outdoor) and in various environments (static/dynamic) have been considered using various algorithms [44–47]. The Fuzzy logic technique can make decisions such as humans in avoiding obstacles in a complex environment, structured and unstructured. There are many variations of the concept of fuzzy logic that allows objects to receive partial membership in uncertain categories obtained using a structure called a fuzzy set. The fuzzy set theory allows defining the behavior of systems by using elements of probability degrees called membership function. The configuration of the general components (fuzzification, rule base, fuzzy inference, and defuzzification) that make up the proposed fuzzy controller is shown in Fig. 4.35. The fuzzification phase occurs from the linguistic variable transformed in each real value's input and outputs of fuzzy sets. The second part is a fuzzy inference that governs the fuzzy logic control process and combines the facts obtained from the defuzzification of the rule base. To transform the subset of output, which is computed by the inference engine, is the primary function of the defuzzification block.

In the proposed method, the type-1 fuzzy logic control algorithm is used for robot path planning- information on how all of the inputs are obtained, as described in the previous chapters. Six input parameters were used. The input parameters required for

Exp.	Path	Convergence	Path Length/ Exec. Time (sec)
1		Best: 901 Mean: 858086 Generation Average Distance Between Individuals Generation Best, Worst, and Mean Scores Generation	ET=8.269551e+01 PL=382
2A			ET=1.654419e+01 PL=1175
2B		Average Distance Between Individuals Best, Worst, and Mean Scores	Convex Hull appl. ET=3.665583e+01 PL=1073
3		Best: 1022 Mean: 507304 Fitness value Average Distance Between Individuals Average Distance Best, Worst, and Mean Scores Generation	ET=2.649140e+02 PL =1068
4A		Best: 1307 Mean: 540451 Fitness value Average Distance Between Individuals Best, Worst, and Mean Scores Generation	ET=3.742653e+01 PL=1084
4B		Average Distance Between Individuals Best, Worst, and Mean Scores	ET=9.262215e+01 PL=1004
5A			ET=8.226215e+01 PL=988
5B			ET=8.226215e+01 PL=1018

Fig. 4.34 Experimental results using GA algorithm in the static environment (2A, 4A, and 5A is the normal environment; 2B, 4B, and 5B is the convex hull applied environment ET: Execution Time; PL: Path Length)

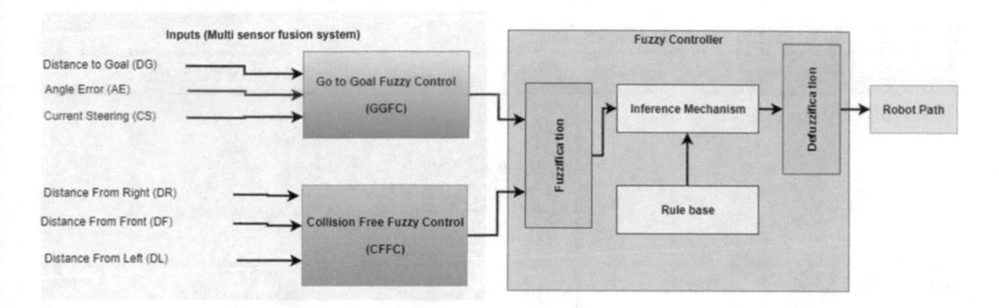

Fig. 4.35 The proposed fuzzy logic approach for mobile robot path planning

this method are shown in Fig. 4.35. These are the angles of the robot measured to the target direction which is always between −180 degrees and 180 degrees, distance from target, distances from obstacles which are distances between the robot and obstacles measured from virtual sensors around the robot, and turn that the robot must avoid the closest obstacle stand for forward, left or right turn respectively. The distances are normalized to be between 0 and 1, multiplied by a constant. The algorithm was evaluated in two stages. In the first stage, the mobile robot aims to act to go to the goal, while in the second stage, the obstacle avoidance action is realized. We designed an expert system for this methodology and created appropriate decision rules for the desired output (path) values. Fuzzy control with different types of membership functions have been designed, and fuzzy control algorithms have been developed for mobile robot behavior control. The details about fuzzy logic components are given in the following sections.

4.2.7.1 Fuzzification

The Fuzzification process comprises a scale of transformation of the fuzzy set convert into suitable linguistic variables. First, the system defines fuzzy variables that correspond to input variables. The linguistic variables used in our application and the corresponding linguistic terms are summarized in Table 4.3—linguistic variables and their corresponding linguistic terms—the MFs consist of one or several types-1 fuzzy sets. The selection of fuzzy sets is based on expert judgment using natural language terms that define fuzzy values that MFs may be triangular, trapezoid, or bell-shaped. These graphically represent a fuzzy set in which the x-axis represents the discourse universe, and the y-axis represents membership degrees in the range [0, 1]. A list of the shapes and characteristics of the existing MFs developed and used is given in Table 4.2.

Depending on the type of problem, and the experience level of the expert, the number of sets and MFs are selected. In our application, the preferred type of MFs is triangular and Gaussian functions. There is no standard design method that can be followed to create an effective solution for the number and structure of member functions. While the behavior of the fuzzy system can be improved by increasing the

Table 4.2 Membership function shapes [45]

Membership functions	Equation of MFs	Shape of function
Triangular MFs	$\mu(x) = \begin{cases} 0, & x < a \\ \frac{x-a}{b-a}, & a \le x < b \\ \frac{c-x}{c-b}, & b \le x \le c \\ 0, & x > c \end{cases}$	
Trapezoidal MFs	$\mu(x) = \begin{cases} 0, & x < a \\ \frac{x-a}{b-a}, & a \le x < b \\ 1, & b \le x \le c \\ \frac{d-x}{d-c}, & c < x \le d \\ 0, & x > d \end{cases}$	
Gaussian MFs	$\mu(x) = exp\left[\frac{-(x-c)^2}{2\sigma^2}\right]$	
S-Shape MFs	$\mu_A(x) = \begin{cases} \frac{x-a}{b-a} & a \le x \le b \\ 1 & b \le x \\ 0 & x \le a \end{cases}$	
Z-Shape MFs	$\mu_A(x) = \begin{cases} 1 & x \le c \\ \frac{d-x}{d-c} & c \le x \le d \\ 0 & x \ge d \end{cases}$	

*Where c and σ, nt the center and width of the graph and, respectively

number of MFs, it can also increase the computational time required for real-time applications and increase the number of rules leading to the formation of complex rules. Considering all these situations, the most appropriate number of MFs and the structure of these functions have been preferred in order to obtain appropriate results against the required inputs.

In fuzzy logic, the degree of membership is a valuation parameter that represents the ownership of a particular event or situation. The membership function represents the fuzzy set $(\widetilde{A})$ is usually donated by μ_A. The representation of membership function distribution for proposed global path planning behavior and the structure of the fuzzy system is graphically illustrated in Fig. 4.36.

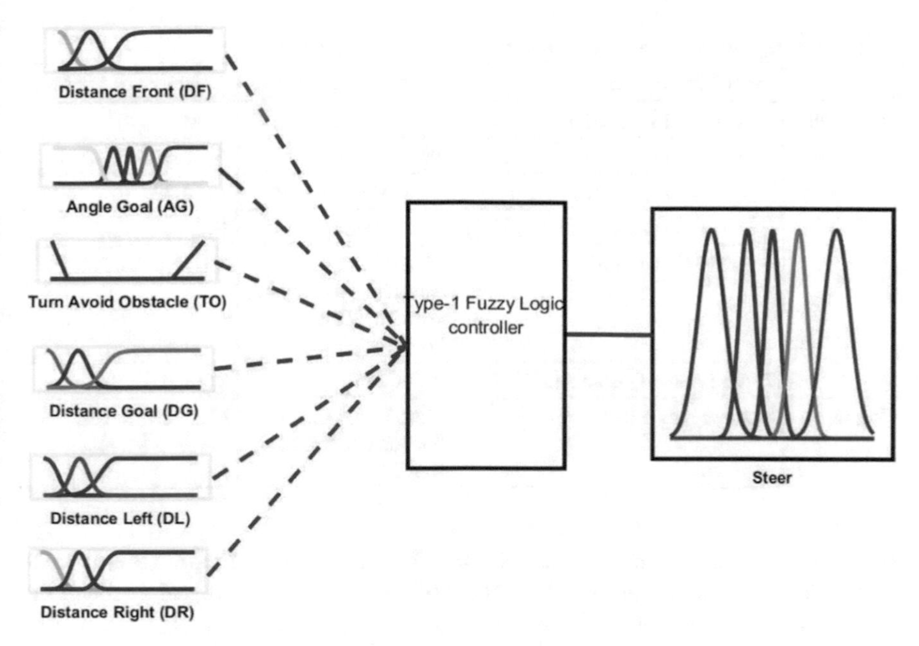

Fig. 4.36 The structure of the fuzzy system: 6 inputs, 1 output, 48 rules

4.2.7.2 Fuzzy Inference Engine

The inference engine is an interface that processes input values regarding specific rules, which are part of the FIS core that drive the system formulated based on human perception and produced fuzzy output sets. The inference engine creates the intermediate stage between the fuzzification and defuzzification of the fuzzy system.

The inference engine consists of rules that usually use logical operators to combine input and output units. These operators, also known as Max-Min operators, take into account the basic methodology and language meaning of AND, OR, NOT. Logical operators receive fuzzy inputs and produce fuzzy outputs. In this phase, the fuzzy logic principle is used to map fuzzy input sets (X1 x…x Xp) that are based on IF-THEN rules, which is interpreted as a fuzzy implication through to fuzzy output set. The desired behavior is defined by a set of linguistic rules. For instance, a type-1 fuzzy (T1F) logic with p inputs (x1 ϵ X1… xp ϵ Xp) and one output (y ϵ Y) with M rules have the following form.

$$R : IF\ x_1\ is\ \widetilde{A}\ and/or\ x_2\ is\ \widetilde{B}\ THEN\ y\ is\ G$$

In these experiments, we used T1F sets and a minimum t-norm operation. To reduce the number of input parameters and make the rule table understandable, the maximum values from the left front diagonal sensor and the left sensor are expressed as left sensor information. The same process has been applied for the right front

Table 4.3 Linguistic variables and their corresponding linguistic terms

	Linguistic variable		Linguistic terms	Abbreviations of term
Inputs MFs	Distances	Right (DR)	Near, Medium, Far	N, M, F
		Left (DL)	Near, Medium, Far	N, M, F
		Front (DF)	Near, Medium, Far	N, M, F
		Distance to Goal (DG)	Near, Medium, Far	N, M, F
	Angle to Goal (AG)		More Negative, Negative, No Change, Positive, More Positive	MN, N, NC, P, MP
	Turn to Avoid Obstacle (TO)		Right, Left	R, L
Output	Steering Angle (SA)		More Left, Left, Forward, Right, More Right	ML, L, F, R, MR

diagonal sensor and right sensors. As a result, DL represents the maximum value of the distance values from the left, and DR represents distance values from the right sensors. In other words, since the sensor value closer to the obstacle will be larger, it does make sense to use this data to perform collision-free path planning. The knowledge bases for each controller consist of 48 rules utilized for the proposed system are presented in Table 4.4.

Table 4.3 provides the meanings of the abbreviations used in Table 4.4 Fuzzy Inference Rules are considered for evaluating the proposed system. The surface generated with the fuzzy inference design is shown in Fig. 4.37.

4.2.7.3 Defuzzification

Defuzzification is the process of producing a measurable crisp result of fuzzy clusters and corresponding membership degrees. It is the process of converting fuzzified inputs to the single crisp output value. There are several well-known defuzzification methods that are the Center of Sums (COS), Centroid of Area (COA)/ Center of gravity (COG) Method, Bisector of Area (BOA), Weighted Average Method, and Mean of Maximum (MOM). In our proposed system, the final output (crisp value) is obtained using the COA [45], which essentially calculates the centroid of the total area representing the fuzzy output set. The result graphs obtained with Type1 Fuzzy Logic control, which we used to find a suitable and optimum path to reach the target from the source, are shown in Fig. 4.38.

The experimental studies reveal that the T1F logic method shows the action of hitting obstacles in sharp turns. The convex hull method has been applied to overcome this situation. After that, the path plan has been completed without collision. The execution time, path length, and virtual sensors data are worked out, and the sensors' data are processed and shown graphically.

Table 4.4 Fuzzy inference rules for the proposed global path

	DL	DF	DR	AG	TO	DT	SA
1		N			LT		MN
2		M			LT		L
3		F			LT		L
4		N			RT		MR
5		M			RT		R
6		F			RT		R
7	N						MR
8	M						R
9			N				MR
10			M				L
11				NA		N	ML
12				MNA		N	ML
13				NC		N	NCD
14				PA		N	MR
15				MPA		N	MR
…							
42				MNA		F	L
43				NA		F	L
44				NC		F	NCD
45				PA		F	R
46				MPA		F	R
47				MNA		M	ML
48				PA		N	MR

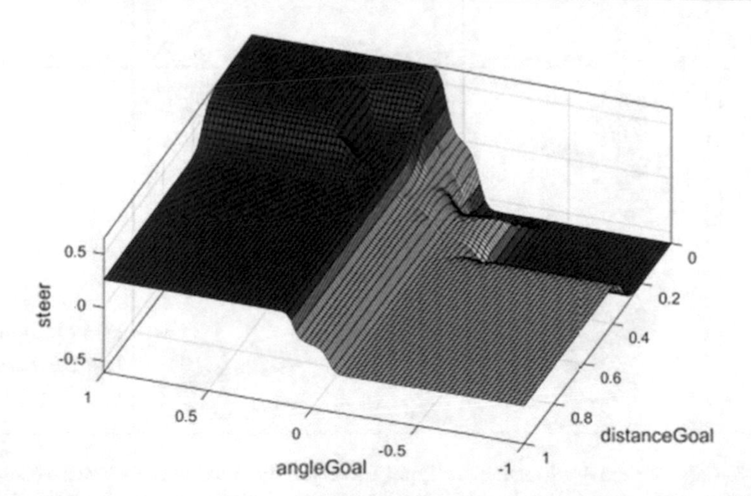

Fig. 4.37 Steering angle control fuzzy surface viewer

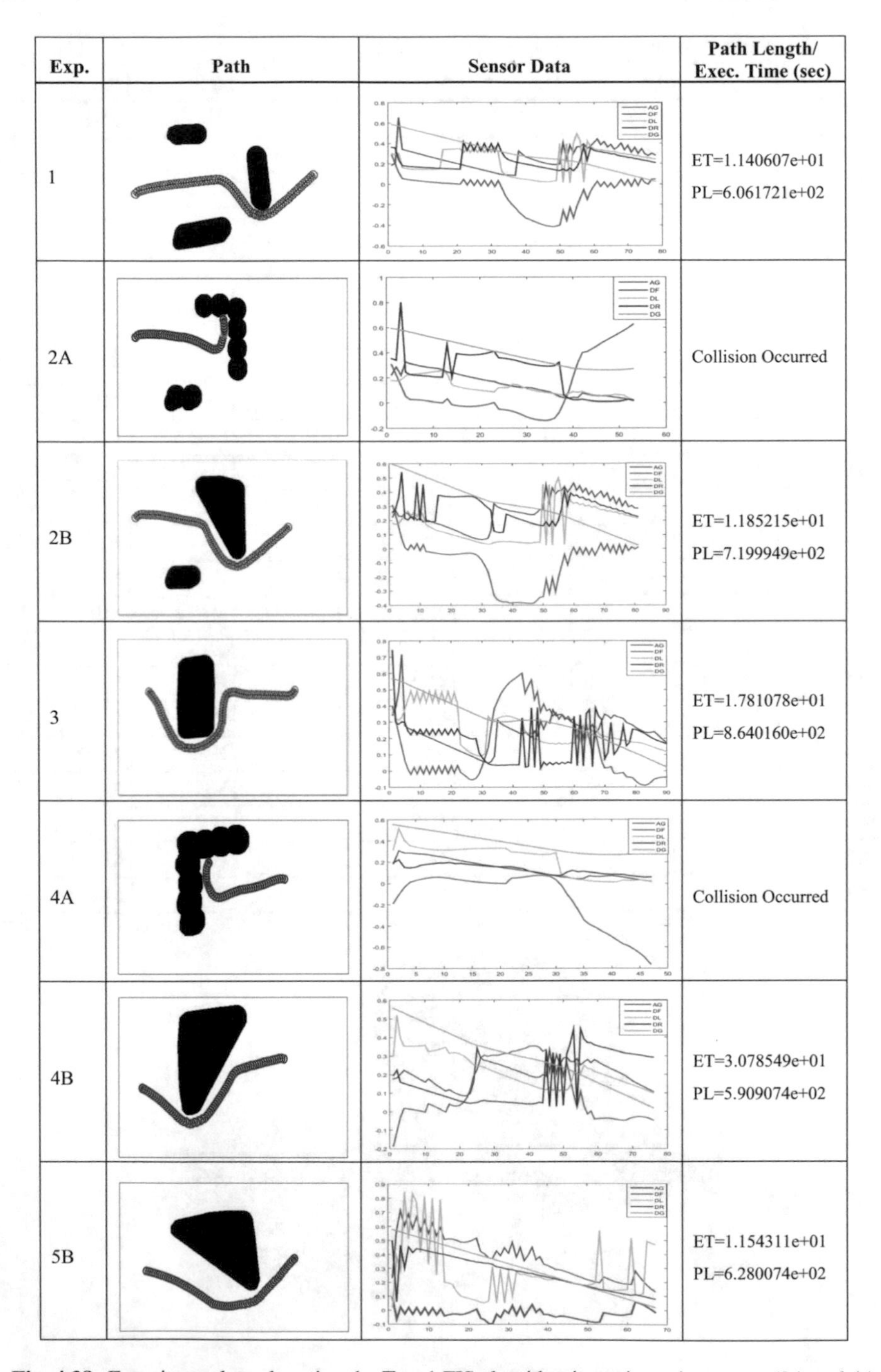

Exp.	Path	Sensor Data	Path Length/ Exec. Time (sec)
1			ET=1.140607e+01 PL=6.061721e+02
2A			Collision Occurred
2B			ET=1.185215e+01 PL=7.199949e+02
3			ET=1.781078e+01 PL=8.640160e+02
4A			Collision Occurred
4B			ET=3.078549e+01 PL=5.909074e+02
5B			ET=1.154311e+01 PL=6.280074e+02

Fig. 4.38 Experimental results using the Type1 FIS algorithm in static environments (2A and 4A is the normal environment; 2B, 4B, and 5B are the convex hull applied environment (ET: Execution Time; PL: Path Length)

4.2.8 *Type 2 Fuzzy Logic*

In this section, first, a brief description of type 2 fuzzy logic is provided, then details of how this method applied in our system are discussed. Next, the application steps and results that form the basis and the design of the T2F model for mobile robot path planning are presented.

In its robustness for controlling nonlinear systems with variation and uncertainties, the T2F method has proven to be a reliable tool for controlling complex systems [48–50]. The presence of uncertainties in a nonlinear system control uses the highest and lowest values of the parameters, extending the T1F method (Fig. 4.39a) to T2F (Fig. 4.39b). Uncertainty is a characteristic of information, which may be incomplete, inaccurate, undefined, inconsistent, and so on. The uncertainty is represented by a region called the footprint of uncertainty (FOU). That is a bounded region that uses an upper and lower T1F membership function.

Type 2 fuzzy logic requires more complex mathematical computation than type-1 fuzzy logic. This complexity is time-consuming in real-time applications. The interval Type-2 Fuzzy Inference System (IT2FIS) has been proposed to minimize this complexity. Despite the advantages of having a third dimension that determines membership for each point in a vertical line of FOU, the computation required for the general case renders T2F unenforceable, because, for general T2F, it is not preferable to compute coverage for all conditions, especially in terms of process complexity. When the number of variables is high, the operation of T2F sets becomes complicated. To solve this problem, Karnik and Mendel [51] were proposed a simple method called interval type 2 fuzzy set (IT2FS).

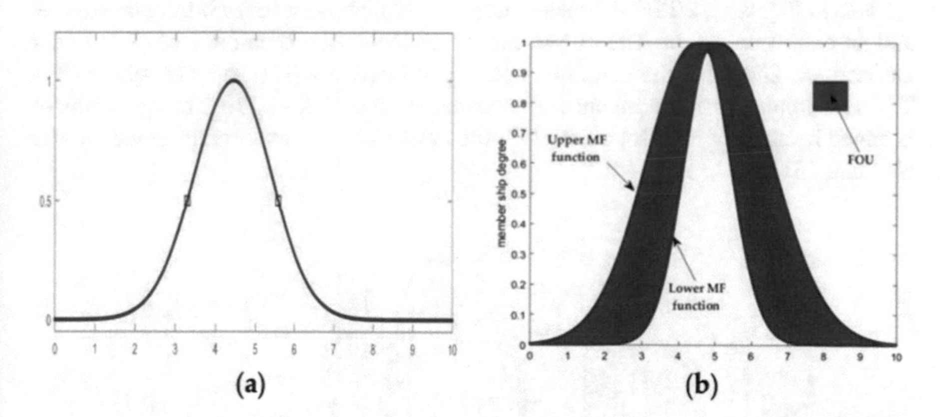

Fig. 4.39 **a** T1F membership function and **b** T2F membership function (IT2FIS). Legend: FOU, the footprint of uncertainty

4.2.8.1 Interval Type-2 Fuzzy Logic System

Interval type-2 is a special uniform function of a T2F membership function (MFs) that second grades take only 1 value. An interval type-2 fuzzy set (IT2FIS) denoted by $\widetilde{A}$, it is expressed in (4.28) or (4.29).

$$\widetilde{A} = \{(x, y), \mu_{\widetilde{A}}(x, y) | \forall_x \epsilon X, \forall_u \epsilon J_x \subseteq [01]\} \tag{4.28}$$

$$\widetilde{A} = \int_{x \in X} \int_{u \in J_x} 1/(x, u) J_x \subseteq [01] \tag{4.29}$$

where $\int \int$ denote the union of all acceptable x and u. An IT2FIS is defined in terms of an upper membership function with $\overline{\mu}_{\widetilde{A}}(x)$ and a lower membership function with $\underline{\mu}_{\widetilde{A}}(x)$. J_x is just the interval of $\left[\overline{\mu}_{\widetilde{A}}(x), \underline{\mu}_{\widetilde{A}}(x)\right]$. The upper membership function (UMF) and lower membership function (LMF) is determined using two T1F MFs which are boundaries of the footprint of Uncertainties (FOU).

Hence, $\mu_{\widetilde{A}}(x, u) = 1, \forall_u \epsilon J_x \subseteq [01]$ is considered as an IT2FIS membership function. The MFs of IT2FIS shape is three dimensional; the third-dimension value is 1 everywhere. It has been found to perform better than T1FS for noisy and well-defined systems in real-time applications that is a specific case where $\mu_{\widetilde{A}}(\boldsymbol{x}, \boldsymbol{u}) = \mathbf{1}$ and the integral form definition is given below (Fig. 4.40)..

A T2F is characterized by IF-THEN rules, where the antecedent and consequent sets are T2F. For designing a T2F controller, it is necessary to know the block structure used with T1F, because the basic blocks used are the same as those used with T1F. As seen in Fig. 4.41, a T2F includes a fuzzifier, a rule base, a fuzzy inference engine, and an output processor. The output processor includes a type-reducer (TR) and a Defuzzifier. The TR is the main distinctive point between T1F and T2F systems. A T1F set outputs or crisp numbers are generated from the TR of T2F. The type reducer is added because of its association with the nature of the membership grades of the elements [51, 52].

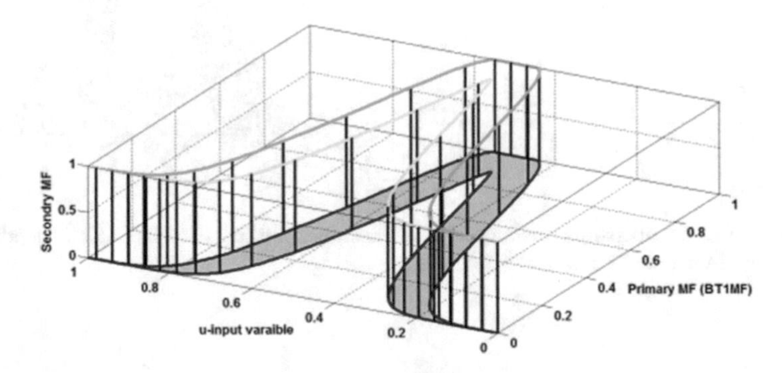

Fig. 4.40 3D Representation of interval type-2 membership function [50]

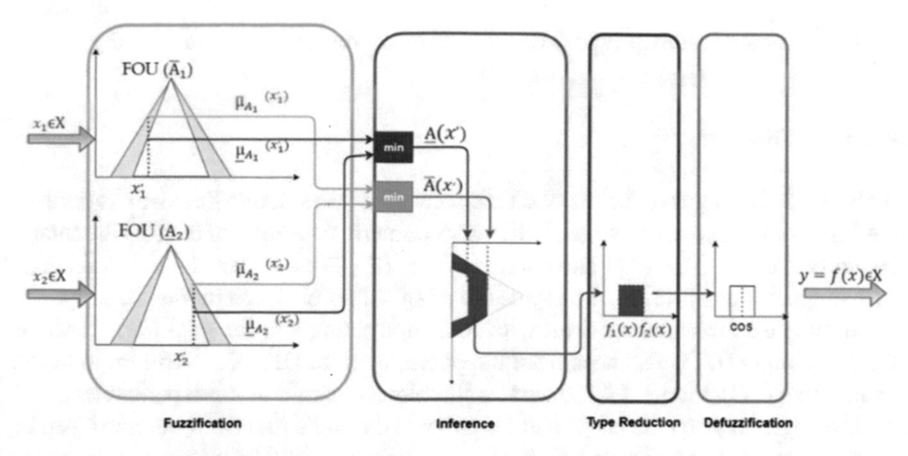

Fig. 4.41 Structure of a type-2 fuzzy logic system

4.2.8.2 Type Reduction

All fuzzy logic systems have to produce a crisp output in order to have a practical application. The single combined T2F set has to be processed with the TR and the defuzzifier. The defuzzifier combines the output sets to obtain a single output using one of the existing RT methods that were proposed by Karnik and Mendel [53, 54]. Several methods commonly utilized for type reduction such as centroid type reduction, height type reduction, and center of the set. In these experiments, a center of sets (cos) type reduction method was used. The mathematical expression of this method is shown in Eq. (4.30).

$$Y_{cos}(x) = [y_l, y_r] = \int_{y^1\in[y_l^1, y_r^1]} \cdots \int_{y^1\in[y_l^M, y_r^M]} \int_{f^1\in[\underline{f}^1, \bar{f}^1]} \cdots \int_{f^M\in[\underline{f}^M, \bar{f}^M]} \Big/ \frac{\sum_{i=1}^{M} f^i y^i}{\sum_{i=1}^{M} f^i} \tag{4.30}$$

This center of sets is completely characterized by its left and right endpoints. The consequent set of the IT2FIS determined those two endpoints (y_l, y_r). If the values of f_i and y_i which are associated with y_l are denoted to f_l^i and y_l^i, respectively, and the values of f_i and y_i which are associated with y_r are denoted to f_r^i and y_r^i respectively, $\underline{f}^i$ and $\bar{f}^i$ are the lower and upper firing degrees of the $i\ th$ rule, and M is the number of fired rules. These points are given in Eqs. (4.31) and (4.32).

$$y_l = \frac{\sum_{i=1}^{M} f_l^i y_l^i}{\sum_{i=1}^{M} f_l^i} \tag{4.31}$$

$$y_r = \frac{\sum_{i=1}^{M} f_r^i y_r^i}{\sum_{i=1}^{M} f_r^i} \tag{4.32}$$

The outputs of interval type-2 fuzzy system are represented with y_l and y_r.

4.2.8.3 Fuzzifier

In this case, the inputs of the fuzzy set are converted into suitable linguistic variables; the MFs consist of one or several T2F sets. A numerical vector x of the fuzzifier maps converts into a T2F set (A). The outputs of the T2F sets are considered a singleton. In a singleton fuzzification, the inputs are crisp values on nonzero membership.

In these experiments, we dealt with four inputs: angle to the goal (θe), distance from the target (DT), distances from the obstacles (DL, DF, DR), and turn to avoid an obstacle (TO). Figure 4.42 illustrates the block diagram of these parameters.

The angle between the target and the robot is the angle that the robot must turn to reach the target. The angle range is between −180 to +180. The distance to the target is normalized between 0 and 1. The distance between the nearest obstacles in the direction in which the robot moves is the distance from the obstacle. This distance is also normalized between 0 and 1 [23]. The movement angle of the robot on the obstacle-free path is taken as the angle of robot direction rotation. There is a single output that measures the turning angle along with the direction. These values are essential when creating the functions of the physical. After the inputs are specified, the selection of the membership functions is performed. The geometry of the MFs is arbitrarily selectable. The MFs used in our application is shown in Fig. 4.43.

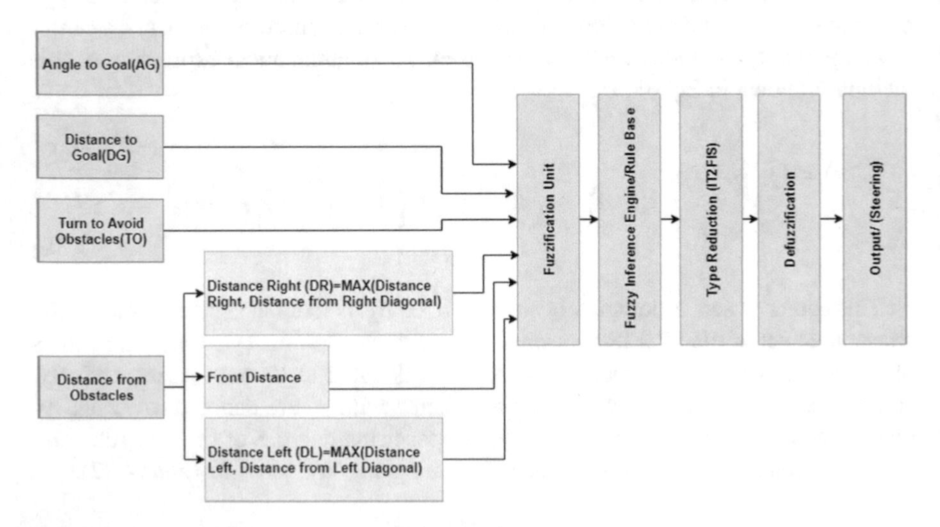

Fig. 4.42 Input parameters of the fuzzy-based robot path planning block diagram

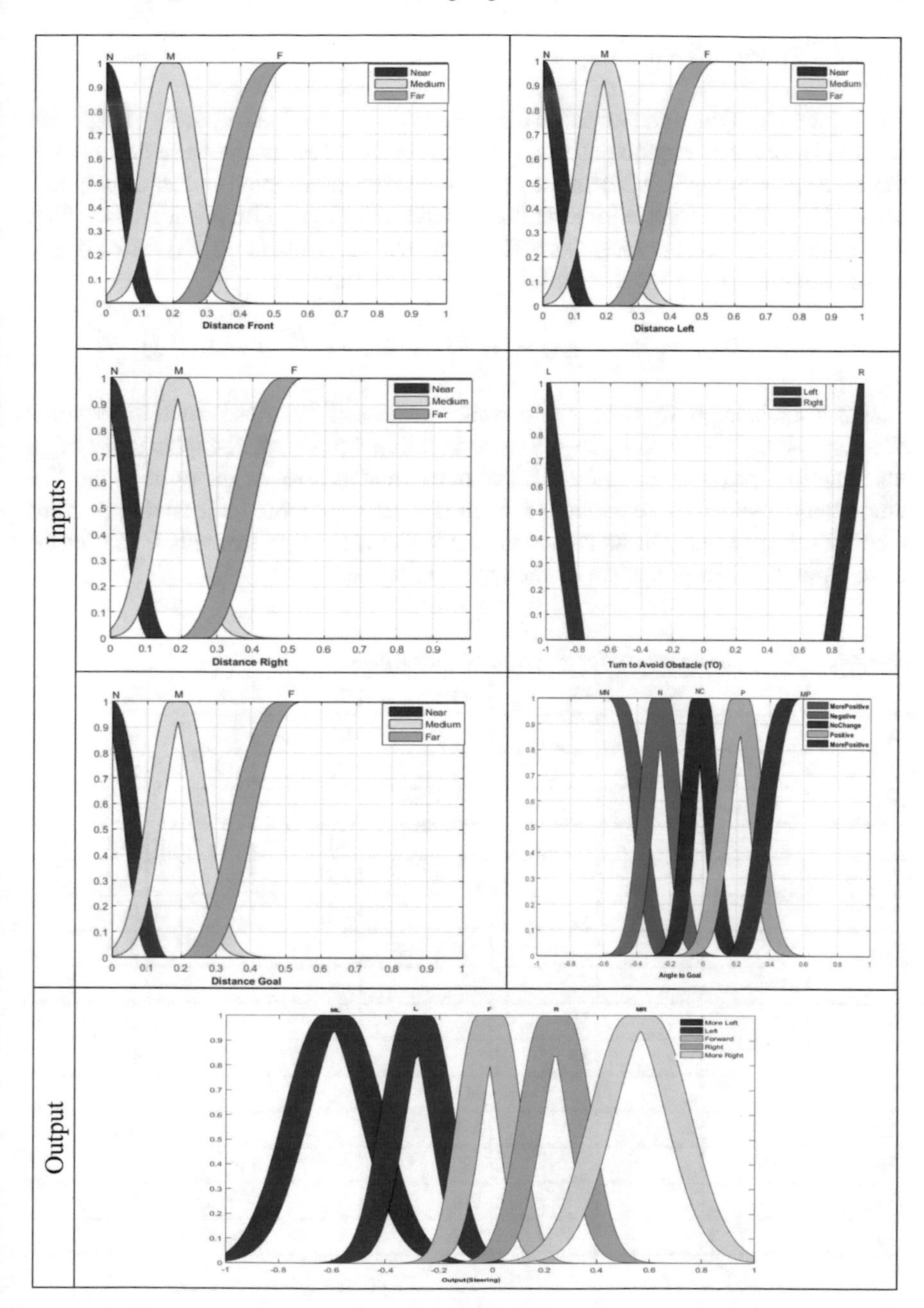

Fig. 4.43 Membership Functions for IT2FIS based path planning

4.2.8.4 Fuzzy Inference Engine

The inference engine is an interface that processes input values according to specific rules and produces fuzzy output sets. That is, it combines or arranges a center between the input and the output. It is necessary to compute the intersection and union of type-2 sets and implement compositions of type-2 relations. A set of linguistic rules defines the desired behavior. For instance, a T2F logic with p inputs ($x_1 \in X_1, \ldots x_p \in X_p$) and one output (y ∈ Y) with M rules has the following form.

$$R^{\ell}: \text{IF } x_1 \text{ is } \widetilde{F}_1^{\ell} \ldots \text{ and } x_p \text{ is } \widetilde{F}_p^{\ell} \text{ THEN } y \text{ is } \widetilde{\mathrm{G}}^{\ell}, \ell = 1 \ldots M$$

The knowledge bases for each controller consist of 48 rules related to the robot Steering angle (SA), which are presented in Table 4.5. Detailed explanations of all abbreviations provided in Table 4.3. The rules created here are the same as the T1F controller explained in the previous section. Therefore, it would be easier to examine the difference between these two controllers. The rule firing strength $F^i(x)$ for the crisp input, the vector is given by the type-1 fuzzy set.

Table 4.5 Fuzzy inference rules for the proposed global path

	DL	DF	DR	AG	TO	DT	SA
1		N			LT		MN
2		M			LT		L
3		F			LT		L
4		N			RT		MR
5		M			RT		R
6		F			RT		R
7	N						MR
8	M						R
9			N				MR
10			M				L
11				NA		N	ML
12				MNA		N	ML
13				NC		N	NCD
14				PA		N	MR
15				MPA		N	MR
…							
44				NC		F	NCD
45				PA		F	R
46				MPA		F	R
47				MNA		M	ML
48				PA		N	MR

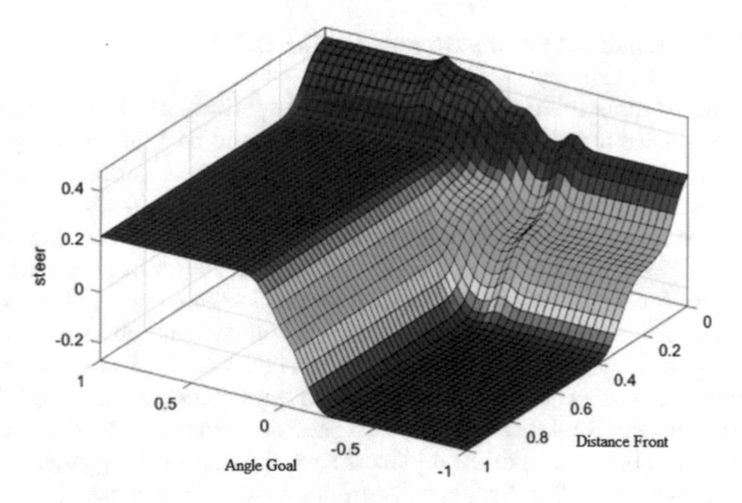

Fig. 4.44 Steering angle control IT2FIS surface viewer

$$F^l(x') = \left[\underline{f}^l(x'), \bar{f}^l(x')\right] \equiv \left[\underline{f}^l, \bar{f}^l\right] \tag{4.33}$$

where $\underline{f}^l$ and $\bar{f}^l$ are the lower and upper firing degrees of the $l\ th$ rule, computed using Eqs. (4.34) and (4.35) in which $*$ represents the t-norm, which is the $prod$ operator in these equations.

$$\underline{f}^l(x') = \underline{\mu}_{\tilde{F}_1^l}(x_1') * \ldots * \underline{\mu}_{\tilde{F}_p^l}(x_p') \tag{4.34}$$

$$\bar{f}^l(x') = \bar{\mu}_{\tilde{F}_1^l}(x_1') * \ldots * \bar{\mu}_{\tilde{F}_1^l}(x_p') \tag{4.35}$$

To evaluate the proposed system, an inference table was created by considering all possibilities. As an example, the fuzzy inference surface constructed by the two inputs and one output parameter is shown in Fig. 4.44.

4.2.8.5 Defuzzifier

The interval fuzzy set $Y_{cos}(x)$ variables obtained from the type reducer have been defuzzified and the average of y_l and y_r have been used to defuzzify the output of an interval singleton type-2 fuzzy logic system. The equation is written as;

$$y(x) = \frac{y_l + y_r}{2} \tag{4.36}$$

4.2.8.6 Experimental Results Obtained Using IT2FIS

To evaluate the applicability of this method, many experiments have been performed. It has been shown that this algorithm can be easily used on a global path planning problem. To thoroughly test the algorithm behavior, we have conducted numerous experiments on different maps. The results are provided in Fig. 4.45.

As shown in these figures, the obstacles in the map are depicted as black regions, and the obstacle-free path is depicted as the white region. The obstacle placed between the robot and the target has a complicated shape and includes the possibility of local minimum problems. The mobile robot probably collides with obstacles with such sharp lines. To overcome this problem, a convex hull technique was applied. With this technique, we observed that the robot's path leads quickly to the target. IT2FIS methods could smoothly manage the path generation process and avoid obstacles. The system has been tested to ensure that the robot monitors sharp angles in cases of disabilities with irregular lines and distortions under two different variable maps. It was first tested under normal conditions (see Fig. 4.45 (Exp. 1, 2A, 4A, 5A)). Secondly, in the new map, by changing the conditions, the boundary lines of the obstacles in the study area were softened, and the gaps were filled with the convex hull method (see Fig. 4.45 (Exp. 3, 2B, 4B, and 5B)). In this case, the path plan was carried out without any collision. The virtual sensor data calculated at each point on the path obtained during path construction are shown on the same graph.

Experiments have been conducted to test the ability of the algorithm to create a safe path in the given map without collisions any obstacles. The primary goal is not the shortest path production; however, finding the safest path in the presence of obstacles of different sizes and shapes is accomplished. The used performance of the path planning algorithms was determined regarding the two following essential criteria that are path length and execution time. As a result of this evaluation, the best algorithm was used in real-time path tracking. General comparison and calculation results have been provided in Chap. 5.

4.3 Real-Time Mobile Robot Path Tracking Process

In this section, we will discuss how to develop a vision-based mobile robot path tracking strategy. This is the third step of the implementation. In this step, the process of following the path obtained in the previous step is performed. The main purpose of the comparison of path planning algorithms is to find the most suitable algorithm according to the determined parameters. Next, the path drawn by this algorithm is followed when going to the path tracking phase. The procedures and explanations followed up to this stage were to form the preliminary steps of the real-time mobile robot path-tracking process. The path coordinates of the most suitable path planning algorithms were used for the robot path tracking process, which is considered as the second stage. Since it is understood that the most suitable planner is IT2FIS from the path planning algorithms, the path coordinates generated by this algorithm

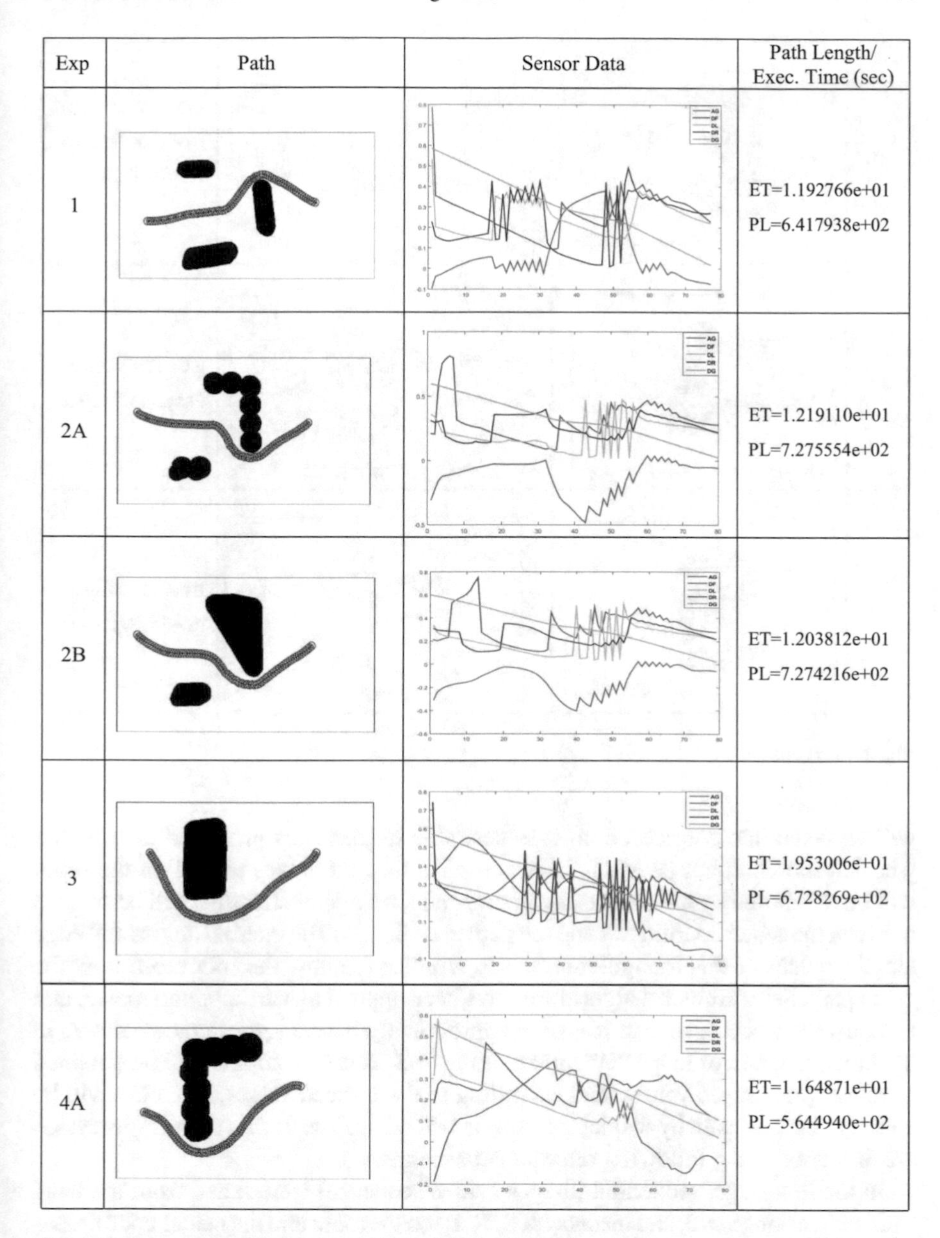

Exp	Path	Sensor Data	Path Length/ Exec. Time (sec)
1			ET=1.192766e+01 PL=6.417938e+02
2A			ET=1.219110e+01 PL=7.275554e+02
2B			ET=1.203812e+01 PL=7.274216e+02
3			ET=1.953006e+01 PL=6.728269e+02
4A			ET=1.164871e+01 PL=5.644940e+02

Fig. 4.45 Path planning experimental results using IT2FIS algorithm in static environments. (1, 2A, 4A, 5A) are normal environments and (3, 2B, 4B, 5B) is the convex hull applied environments (Legend: ET: Execution Time; PL: Path Length)

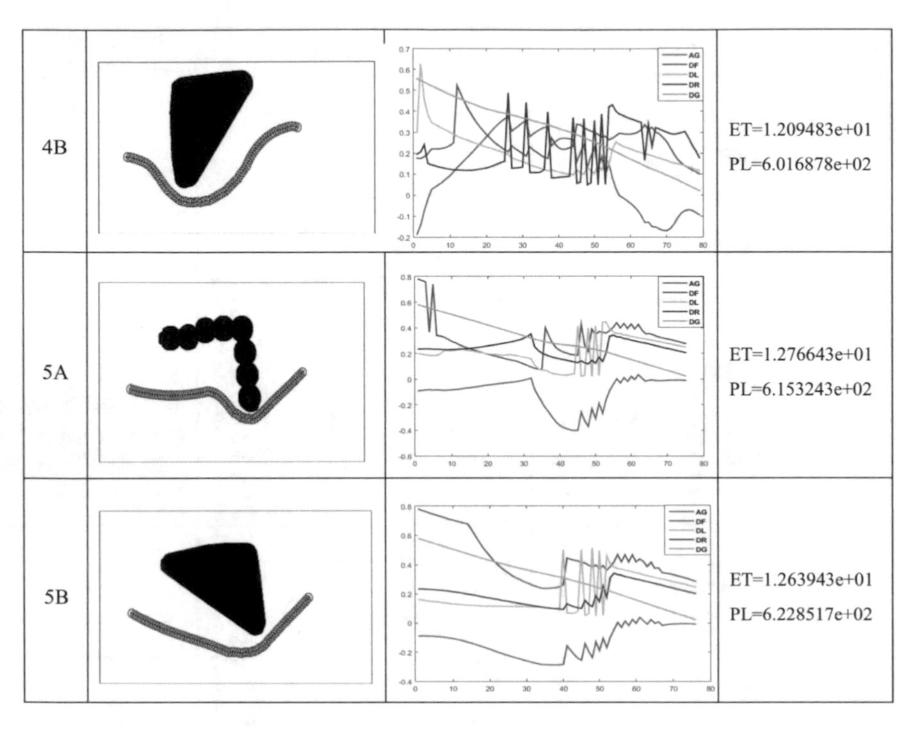

Fig. 4.45 (continued)

will be taken into consideration. It is desirable to track this produced path with a wheeled mobile robot (WMR). To accomplish this, the labels placed on the robot will be monitored by the overhead camera, and a triangular structure will be formed between the center coordinates and the path coordinates. The internal angles and edge lengths values of this triangular structure will use to allow the robot to follow the given path. Several control algorithms have been applied to use these parameters and to adjust the wheel speeds of the robot appropriately. These algorithms are shown in the block diagram of LabVIEW, given in Fig. 4.2. Suitable outputs will be obtained as robot wheel speed values, and according to these speed values, the robot will be able to follow the path by displaying certain behaviors. The details of these processes are discussed in the following relevant subsections.

In the first stage, sequential images (video sequence) were taken from the head camera (working area), and the labels (L, R, F) on the robot and the center coordinates of the target object were obtained. For this purpose, the color pattern matching method was used. A triangular structure was formed between the center coordinates of the obtained templates and the target point. The edge lengths and base angles of this triangle were used as the controller input, and the robot followed the specified path.

Four different controllers were used for this path following procedures. Two of these controllers (Gauss and Dec. tree) have been used in previous studies [55–57]. Two newly developed controllers (Type-1 and Type-2 fuzzy) use the data obtained

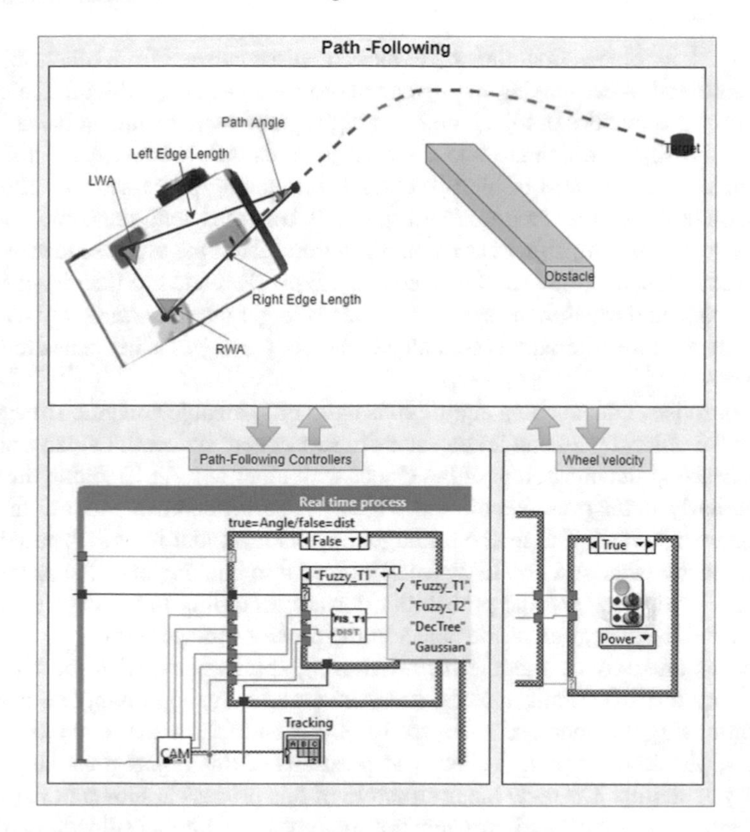

Fig. 4.46 The general perspective of the real-time path tracking control process

from the triangular kinematic diagram as input, allowing the robot to track the path obtained. The LabVIEW block structure developed for real-time path tracking is shown in Fig. 4.2. After the input value (angle or length) for the controller is selected from the interface, the path-tracking parameters are created according to this input value. The control algorithms that have been used to adjust the wheel speeds of the robot appropriately are shown in the block diagram of the LabVIEW in Fig. 4.46. Suitable outputs will be obtained as robot wheel speed values, and according to these speed values, the robot will be able to follow the path by displaying certain behaviors. The general overview of these control processes is given below. Details of these transactions are discussed under the relevant headings under Sect. 4.3.3.

4.3.1 Object Tracking

Object Tracking is a standard function used in computer vision. Accurately identifying and monitoring objects is an important topic that needs to be analyzed. Being able to track an object, first, the object must be distinguished from other objects in the background. After making improvements to the image according to the camera and light characteristics used, the object is highlighted by focusing on the color and shape of the object, which are two essential features that can reveal the difference of the object. The success of high-level decision-making processes will ultimately depend on how well the foreground is separated from the background and how well the foreground moving objects are monitored throughout the video sequence. Next, it is checked whether the area that specifies the pixels related to the object has the desired shape and whether the area it occupies is larger than a particular pixel value. According to these criteria, it is concluded whether the object is in the desired shape and color.

The purpose of the tracking algorithm is to identify an object in video images and identify the object by tracing its trajectory in subsequent sequential video frames or by emphasizing the interaction of the object with other objects. Tracking the object very similarly in the presence of other objects is performed using algorithms such as template matching. The search for target objects in similar images is based on an average shift method in LabVIEW, which is useful in tracking even the presence of other similar objects, and the probability density according to the object's current position and the histogram of the object in the previous image frame.

There is an auxiliary module for monitoring applications using the NI Vision module in LabVIEW. Path tracking was performed by using a template matching algorithm using this module. The steps to obtain and follow the templates we use for the application are given below. The positions of this object must be detected correctly at startup. The code block structure of this process is shown in Fig. 4.47.

The information and block structure for implementing this algorithm in LabVIEW are using the IMAQ library. This was chosen for continuous tracking of the coordinates of the labels being followed and to be able to work independently of the background of the image. It is appropriate to use a template matching algorithm to track an object. In the proposed system, when following the labels placed on the robot, differentiation from the first template may be seen depending on the direction and rotation of the robot.

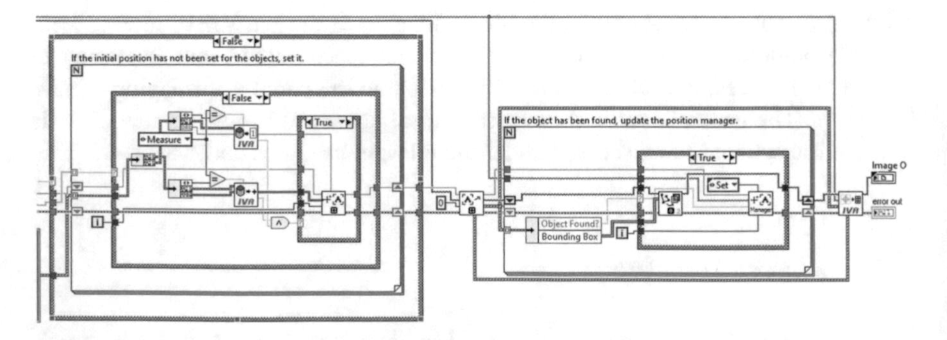

Fig. 4.47 Object tracking block diagram

4.3.2 *Kinematic Analysis of a Mobile Robot*

This section will provide information about the traditional kinematics of mobile robot and motion planning. As it is known, a traveling robot with a differential drive has a non-holonomic structure because it has differential constraints that cannot be fully integrated. The system in which the robot cannot move sideways and moves on the principle of rotating wheels is expressed by the term non-holonomic [58]. In order to make a smooth curved path plan for a traveling robot, the kinematic model of the non-holonomic traveling robot with a differential drive must first be obtained. To construct the kinematic equations of the WMR, it is necessary to define some parameters. These parameters are generated considering Fig. 4.48. This figure illustrates a kinematic and dynamic model of a two-wheeled mobile robot with a non-holonomic differential drive.

In Fig. 4.48, L represents the width between the left and right wheels, R is the radius of the wheels, and C is the center of mass of the mobile robot. The field of the navigation landmark environment is shown with (O: Origin, X, Y), and the moving axis of the mobile robot is shown with (θ, x, y). θ is the angle of rotation representing the orientation of the robot about an axis (O, X). The initial posture of the mobile robot is denoted by three parameters (x, y, θ), as given in (4.37).

$$q = [x, y, \theta]^T \tag{4.37}$$

Equation (4.38) is written for non-holonomic constraints.

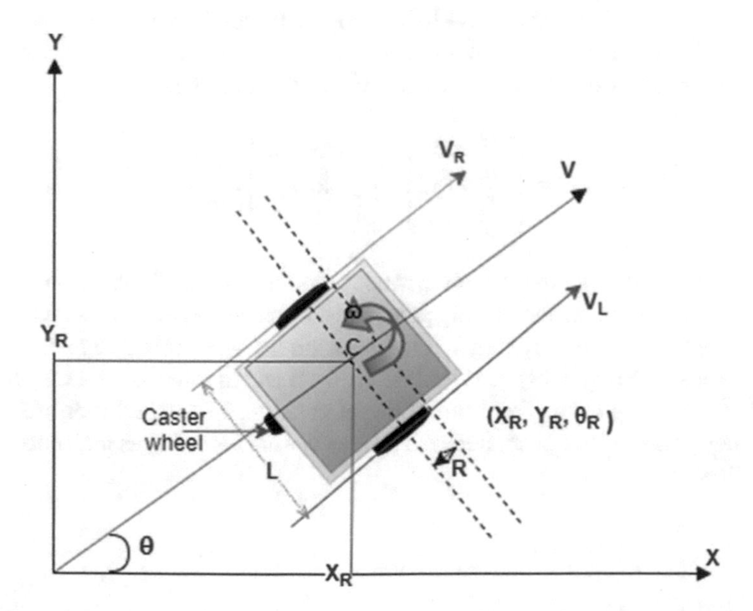

Fig. 4.48 A kinematic and dynamic model of the non-holonomic differential drive two-wheeled mobile robot

$$\dot{y}\cos\theta - \dot{x}\sin\theta = 0 \tag{4.38}$$

The relationship between linear speed and angular velocity of the wheels is explained by the following equations.

$$V = \omega.R \tag{4.39}$$

$$V_R = \omega_R.R \tag{4.40}$$

$$V_L = \omega_L.R \tag{4.41}$$

$$\omega = \frac{V_R - V_L}{L} \tag{4.42}$$

$$V = \frac{V_R + V_L}{2} \tag{4.43}$$

$$\begin{bmatrix} V \\ \omega \end{bmatrix} = \begin{bmatrix} \frac{1}{2} & \frac{1}{2} \\ \frac{1}{L} & \frac{1}{L} \end{bmatrix} \begin{bmatrix} V_R \\ V_L \end{bmatrix} \tag{4.44}$$

In Fig. 4.48, L shows the width between the left and right wheels, R is the radius of the wheels, and C is the center of mass of the mobile robot. The field of the navigation landmark environment is shown with (O: Origin, X, Y), and the moving axis of the mobile robot is shown with (θ, x, y). θ is the angle of rotation representing the orientation of the robot about an axis (O, X). The initial posture of the mobile robot is denoted by three parameters (x, y, θ), as given in (4.45).

$$\dot{q} = \begin{bmatrix} \frac{dx}{dt} = \dot{x} \\ \frac{dy}{dt} = \dot{y} \\ \frac{d\theta}{dt} = \dot{\theta} \end{bmatrix} = \begin{bmatrix} \cos\theta & 0 \\ \sin\theta & 0 \\ 0 & 1 \end{bmatrix} \begin{bmatrix} V \\ \omega \end{bmatrix} \tag{4.45}$$

These equations apply to the mobile robot's motion control. To plan the movement of a mobile robot between the starting point and the desired target point, specific nodes must be referenced. In this system, circles, arcs, curves, and straight lines can be used to form orbits (see Fig. 4.49). It is first determined which joints the robots will pass to reach the target position from their current position. The trajectories are formed so that they pass over the given nodes. These nodes are recorded as path coordinates to be followed.

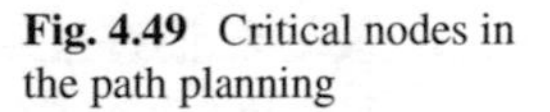
Fig. 4.49 Critical nodes in the path planning

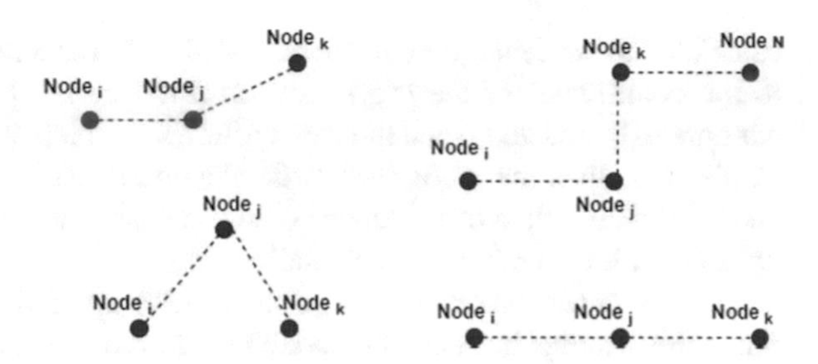

4.3.3 Proposed Control Architecture

In this section, a kinematic control approach is proposed to run and keep up the mobile robot along the generated path. Lego Mindstorms [59, 60] differential drive mobile robot, which consists of the two front driving wheels and one caster wheel for carrying the chassis was used for this purpose. This robot is mounted two separate motors for driving the wheels and control the motion and orientation of the robot. The structure of the proposed system is illustrated in Fig. 4.50. To conduct the experiments, the parameters presented in Fig. 4.50 were utilized.

In Fig. 4.50, L (l_x, l_y), F (f_x, f_y), and R (r_x, r_y) refer to the center coordinates of the labels (L, F, R) placed on the robot top. L and R are placed on the robot coinciding with the wheels. F is used to indicate the direction of the robot. T (t_x, t_y) represents the target coordinate and C (c_x, c_y) represents the robot's central points, which is

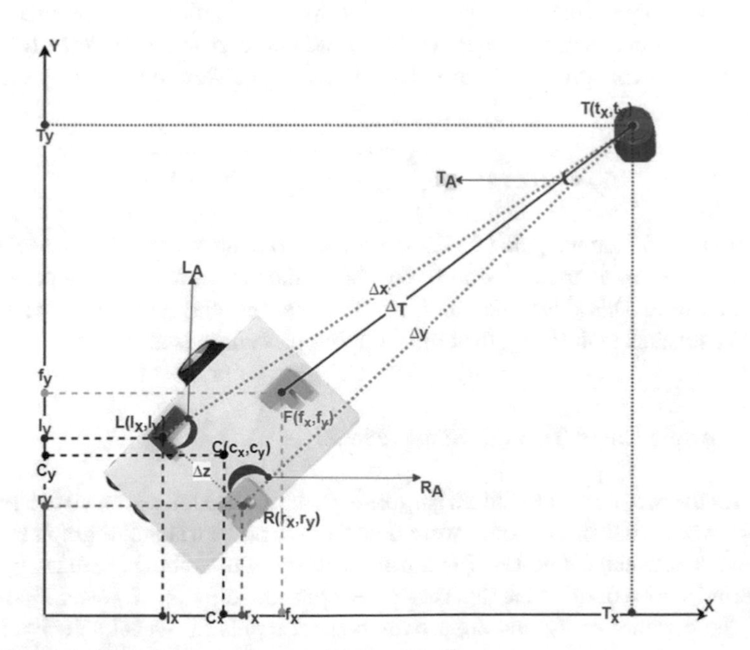

Fig. 4.50 Triangle-based proposed positioning model scheme

calculated using center coordinate values of L, F and R. The target coordinates refer to the coordinates of the previously obtained global path. This situation will be continuously updated when the robot follows the path, and the final coordinate will be reached when the robot reaches the destination. If the distance value between F and T is smaller than the distance between L and R and the new target coordinate is updated with ten path nodes hospitality.

Several image processing steps have been applied for obtaining these labels. Since this process has been elaborated on in the previous sections. A graph-based triangle shape was created between these labels. This triangle is continuously updated according to a threshold value between the wheel labels and an intermediate target on the formed path. The right-left and target internal angles of this structure is denoted with R_A, L_A, and T_A and the lengths between these nodes (L, R, T) is denoted with $\Delta_x, \Delta_y, and\ \Delta_z$ respectively. Other details of this structure are given in the following sections.

4.3.3.1 Distance-Based Triangle Shape Model

To create the proposed structure, the graph obtained in Fig. 4.50 was used. Four vertices (F, R, L, and T) were used in this graph. Each edge of this graph has a weight, which is used to control the robot when navigating. The edge lengths of this structure were obtained by a number of trigonometric calculation methods. Δ_x, Δ_y, Δ_z, and Δ_T are indicate the weight of the edges between nodes. The distance between the robot and the target position is indicated by Δ_T. These distances were calculated using the Euclidean theorem. Equation (4.46) was used to obtain the left edge length of the proposed triangular structure. The right lengths were obtained by the same theorem.

$$\Delta_x = d(L, P) = \sqrt{(P_x - L_x)^2 + \left(P_y - L_y\right)^2} \tag{4.46}$$

Right and left side weights (Δ_x, Δ_y) are used to control the WMR wheel velocities. The edge Δ_T is used to stop the mobile robot when it reaches a pre-defined threshold value. This threshold value is calculated by using the Δ_z edge weight. When DT reaches to 4/3* Δ_z, then the stopping procedure is activated.

4.3.3.2 Angle-Based Triangle Shape Model

To control the mobile robot with a triangle shape-based model, as shown in Fig. 4.49, first, the coordinates of the nodes were determined, and a triangular graph structure was created between the nodes. The motion control of the robot is realized by using the internal angles of this triangle. These are represented by R_A, L_A, and T_A, respectively. The parameters R_A and L_A are the control inputs for wheel velocity. On the other hand, the T_A parameter is used to stop the mobile robot when it reaches a predefined threshold. To regulate input parameters for the controller, acquired angle data are compared dynamically. Whenever the robot moves, the angle values of tracked objects are updated periodically by using the coordinate values in the 2D (x-y) space. Several trigonometric calculation methods obtained the internal angles of this structure. The interior angle values were calculated using the Law of Cosine [61] theorem given in Eq. (4.47).

$$R_A = \mathrm{a}\cos\left(\frac{\Delta_{\mathrm{x}}^2 - \Delta_{\mathrm{y}}^2 - \Delta_{\mathrm{z}}^2}{-2 * \Delta_{\mathrm{y}} * \Delta_{\mathrm{z}}}\right) \tag{4.47}$$

$$\mathrm{Ang}_{R_A} = R_A * \frac{180}{\mathrm{pi}} \tag{4.48}$$

Δ_x, Δ_y, and Δ_z are distance vectors between the centroid coordinates of the detected objects. The $'R_A'$ is the radian value of the related angle and $'\mathrm{Ang}_{R_A}'$ is the conversion of this radian value to the degree value. Velocity values for both wheels are computed according to these angle values for the triangle-based approach. The same theorem is used to obtain the L_A and T_A.

4.3.4 *Proposed Control Algorithms*

For global path planning and path tracking applications, mobile robots must have the ability to reach the destination from the starting point within its workspace with avoiding any collision which can cause harm to the robot. In this work, a hybrid control architecture has been developed to meet these requirements and to control the mobile robot. This architecture is combined with the idea of minimizing the systematic or non-systematic error of the rapid response of reactive architecture by its strong vision-based planning ability. Four different control algorithms have been utilized on the proposed kinematic control approach in order to realize the motion control of the wheeled mobile robot. The details of these algorithms are discussed next.

4.3.4.1 Gaussian Control Algorithm

The Gaussian (normal) distribution that has many mathematically advantageous characteristics is used across engineering disciplines. The single-dimensional general Gaussian function is given in Eq. (4.49).

$$f_G(x) = \frac{1}{\sigma\sqrt{2\pi}} e^{-\frac{(x-\mu)^2}{2\sigma^2}} \tag{4.49}$$

This equation is distributed according to the parameters of the mean (μ) and standard deviation (σ). The input parameter (x) in this equation is the difference between Δ_x, Δ_y, or R_A, L_A, which are graph weight (edge length) values and interior angle of the proposed visual-based control scheme as detailed previously. The following equations were used to calculate the input parameter (x).

$$X_{(Edge)} = \left| \frac{\Delta_y - \Delta_x}{\lambda} \right| \tag{4.49}$$

$$X_{(Angle)} = \left| \frac{L_A - R_A}{\varphi} \right| \tag{4.50}$$

In these equations, $X_{(Edge)}$ and $X_{(Angle)}$ represent the absolute difference value for distance and angle approaches, and λ and φ are fixed constants that are found as empirically. These constants are used as smoothing factors to adjust the amplitude of the $X_{A/E}$ output; this process smooths wheel reactions and facilitates the calculation of velocity parameters. To calculate the effect of the Gaussian function on wheel speeds and models, the go-to-goal behavior of the mobile robot, Eq. (4.51), was used.

$$V_C = V_{max} * (1 - f_G) \tag{4.51}$$

where V_C is used as the wheel speed coefficient. V_{max} is the maximum speed of the wheel and f_G is a Gaussian function. The difference in $(1 - f_G)$ provides inverse effects on the wheels as a key approach in adjusting the wheel positions to compensate for the robot position. Equations obtained and used to calculate the final value of the wheels are given below.

Distance – Based Gaussian Control(GGC) Angle – Based Gaussian Control (AGC)

$$\boldsymbol{V_L} = \begin{cases} \boldsymbol{V_{max} * \tau + V_C,\ \Delta_x < \Delta_y} \\ \boldsymbol{V_{max} * \tau - V_C,\ \Delta_x > \Delta_y} \end{cases} \quad V_L = \begin{cases} V_{max} * \gamma - V_C,\ R_A < L_A \\ V_{max} * \gamma + V_C,\ R_A > L_A \end{cases} \tag{4.52}$$

$$\boldsymbol{V_R} = \begin{cases} \boldsymbol{V_{max} * \tau + V_C,\ \Delta_x > \Delta_y} \\ \boldsymbol{V_{max} * \tau - V_C,\ \Delta_x < \Delta_y} \end{cases} \quad V_R = \begin{cases} V_{max} * \gamma - V_C,\ R_A > L_A \\ V_{max} * \gamma + V_C,\ R_A < L_A \end{cases} \tag{4.53}$$

$$\boldsymbol{V_{L,R} = V_{max} * \tau + V_C,\quad \Delta_x \cong \Delta_y} \quad \begin{matrix} V_{R,} = \mathrm{Vmax} * \gamma + V_C,\quad \mathrm{Thr} \le T_A \\ V_L = 0\ \&\ V_R = 0 \text{ iff } T_A \ge 60 \end{matrix} \tag{4.54}$$

V_L and V_R correspond to the left and right wheel velocity of the mobile robot, respectively. The $'\tau'$ and $'\gamma'$ are used as a constant scaling factor. The wheel velocities of the robot were calculated by comparing the side lengths and base angles of the proposed structure. According to this comparison, the right wheel velocity increases while the left one decreases, i.e., it is affected in the opposite direction. The graph of the results obtained by performing this algorithm is given below for the results of 5 different experiments. The experiments obtained using Gauss-based path tracking are given in Fig. 4.51.

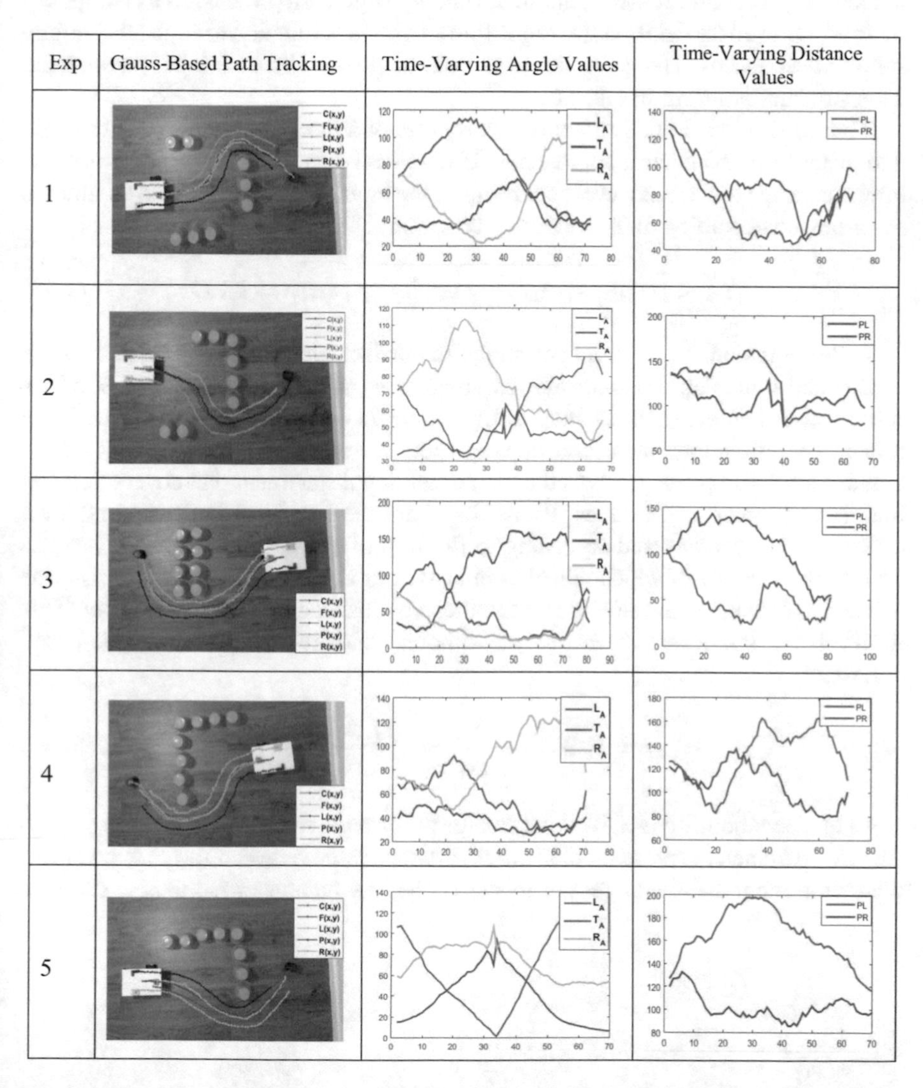

Fig. 4.51 Experimental results using a gauss-based path-tracking algorithm in static environments. (Legend: L_A: left wheel angle; R_A: right wheel angle; T_A: target angle)

4.3.4.2 Decision Tree Control Algorithm

A decision tree is an essential tool for determining the most suitable output from a range of candidate possibilities. This technique is an appropriate solution to automate a decision procedure. It can be applied to many engineering fields. In the proposed method, we used the decision tree algorithm as a control function to adjust the mobile robot wheel speeds. The controller uses both angle and distance input parameters. This structure is shown in Fig. 4.52.

The $\Delta_{D/A}$ given in this structure has represented the input value for both angle and distance-based control approaches. If the given conditions are satisfactory, the branches may vary at each level according to the input parameters. These conditions are arranged according to the following equation.

$$\Delta_{\mathrm{n}} < \left|\Delta_{\mathrm{A/D}}\right| < \Delta_{\mathrm{m}} ? \alpha = \alpha_{\mathrm{n}}, \beta = \beta_{\mathrm{n}} : \alpha, \beta = \mathrm{NRC} \tag{4.55}$$

In this equation, A_n and A_m correspond to the distance values for a related condition. α and β are denoted controller parameters. α_n and β_n are new α and β values if the required statement is satisfied. If the statement does not satisfy in the available related condition, then the α and β parameters are searched in NRC (Next Range Condition). NRC corresponds to the next conditional statement that covers the next bottom and upper angle values. Branching from the first level to the second level of the tree has been altered according to the sign of $\Delta_{A/D}$ parameter. This process defines the velocity of which wheel is increased or decreased. Equation (4.57) was utilized to calculate the velocity limit that can be reached by the mobile robot. This limit value is significant to determine upper bound velocity for the decision tree-based controller.

$$V_{max} = V_c + V_c * \frac{\Delta_{T_n}}{\Delta_T} = V_c\left(\frac{\Delta_T + \Delta_{T_n}}{\Delta_T}\right) \tag{4.56}$$

In this equation, Vc is a fixed parameter to form the initial value of the velocity. The Δ_T parameter represents the initial distance value between the first positions where the robot and the target are located. The Δ_{T_n} parameter is the new distance

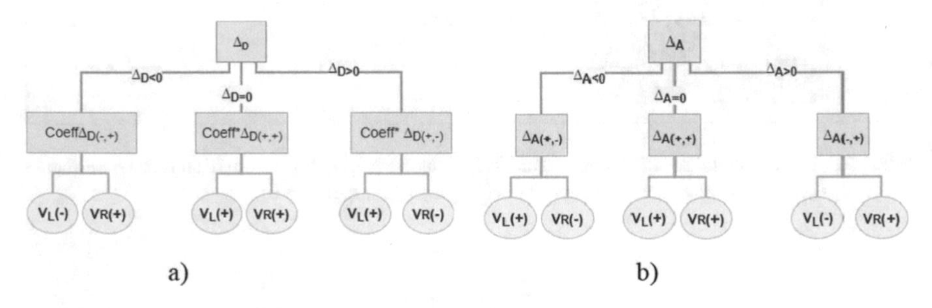

Fig. 4.52 Decision tree structure designed for robot control **a** Distance-based, **b** Angle-based

value, which is the current distance between the robot and the target positions. In the table below, using the decision tree structure, both angle-based and distance-based wheel speed calculation formulas are given.

Distance – Based Dec. Tree control **Angle – Based Dec. Tree control**

$$\Delta_D = \frac{\Delta_1 + \Delta_2}{2} \quad A_A = \frac{L_A + R_A}{2} \tag{4.57}$$

$$\gamma = \frac{\alpha + \beta}{2} \quad \gamma = \frac{\alpha + \beta}{2} \tag{4.58}$$

$$V_L = V_{max} * \left(\frac{\Delta_T}{\Delta_1}\right) * \alpha \quad V_L = (V_{max} + R_A) * \alpha \mp \sqrt{(A_d + 1) * \frac{1}{T_A + 1}} \tag{4.59}$$

$$V_R = V_{max} * \left(\frac{\Delta_T}{\Delta_2}\right) * \beta \quad V_R = (V_{max} + L_A) * \beta \mp \sqrt{(A_d + 1) * \frac{1}{T_A + 1}} \tag{4.60}$$

$$V_{R,L} = V_{max} * \left(\frac{\Delta_T}{\Delta_{1,2}}\right) * \gamma \quad V_{R,L} = (V_{max} + A_A) * \gamma + \sqrt{(A_d + 1) * \frac{1}{T_A + 1}} \tag{4.61}$$

The average A_A and Δ_D is calculated by using Eq. (4.58). These parameters are used when the L_A and R_A values have a value in the predefined difference threshold values. Similarly, the average (γ) of the controller parameters (α and β) is calculated by using the Eq. (4.59). This γ parameter is only used when the angle (L_A and R_A) and distance (Δ_1 and Δ_2) values have a value in the pre-defined difference threshold values as well. The values of wheel velocity are calculated using the following Eqs. (4.60)–(4.62).

V_L and V_R are the velocity values for the mobile robot. R_A , L_A and T_A parameters are the interior angle values of the triangle structure. The angle T_A which is located at the target position is used to affect the final velocity values to ensure a balance between velocity values (by increasing or decreasing). Δ_1, Δ_2 and Δ_T parameters are the distance values of the triangle structure. The distance Δ_T, which is located between the target position and robot front label, is used to affect the final velocity values to ensure a balance between velocity values. The results obtained by performing a Decision tree control algorithm is given in Fig. 4.53 for five different experiments.

4.3.4.3 Type 1 Fuzzy Logic Control

Type-1 fuzzy logic, which is one of the control methods developed for mobile robot path tracking applications. This is a vision-based method that has been used for global path tracking in static environments. The parameters to be used as control

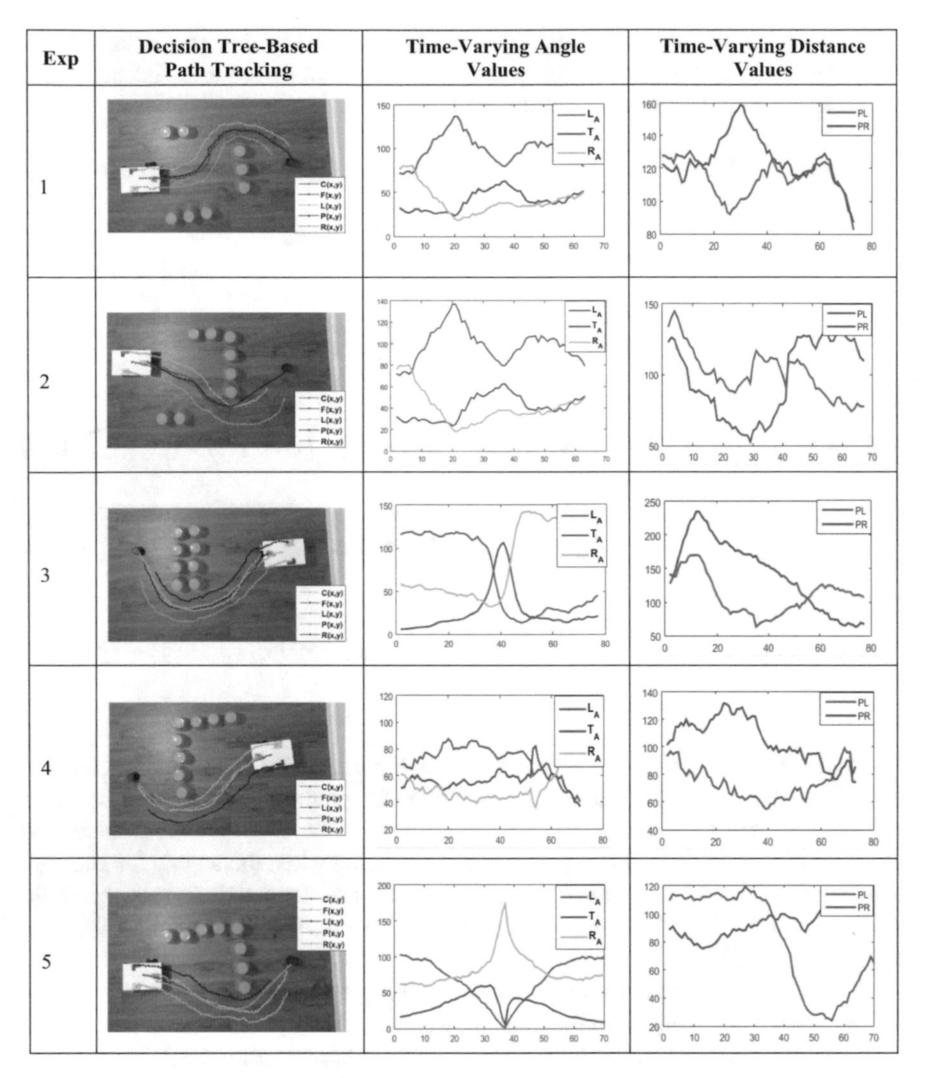

Fig. 4.53 Experimental results using decision tree-based path tracking algorithm in static environments. (Legend: LA: left wheel angle; RA: right wheel angle; TA: target angle)

inputs are the edge lengths of the proposed triangular-based positioning scheme (see Fig. 4.50). In our previous studies, we used angle values as a control input. In this work, the edge lengths are taken as input parameters, and the performance of the results obtained with angle values are compared. These inputs were used to adjust the wheel speed of the mobile robot to perform predetermined path tracking.

The accuracy of path tracking is mainly dependent on the control strategy and real-time controller performance. The combination of vision technology and fuzzy logic control techniques can improve the real-time control performance of a mobile

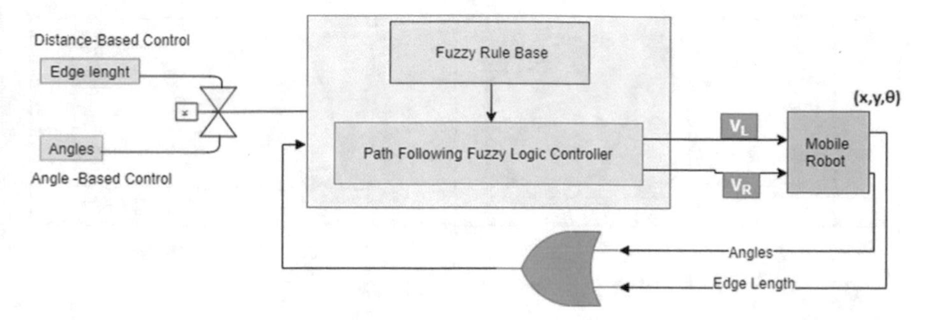

Fig. 4.54 Block diagram of the path-tracking type-1 fuzzy control system

robot. An expected error occurs between the current robot position and the desired path due to mechanical shocks caused by sliding between the moving robot wheels and the ground or systematic and nonsystematic errors. The method developed for visual-based robot navigation is aimed to reduce the error rate to be obtained from real-time robot path tracking applications using soft calculation techniques. The error is measured by the distance of the robot to the specified path.

The structure has two FLC controllers that are used to produce wheel speed. The first controller is distance-based, and the second controller is angle-based triangle control techniques. This technique is shown graphically in Fig. 4.54.

Fuzzy logic control has three procedures: fuzzification, rule-based structure, and defuzzification. In this section, the used membership functions, the rule table, and the graphical surface will be explained.

A designed expert system for this methodology has two inputs, two outputs. Appropriate decision rules for the desired output (wheel speed values) have been determined. Fuzzy control designs with different MFs have been designed, and fuzzy control algorithms have been developed for mobile robot behavior control. Depending on the type of problem to be solved, and the experience of the expert, the number of sets and MFs are selected. There is no standard design method that can be followed to create an effective solution for the number and structure of member functions. While the behavior of the fuzzy system can be improved by increasing the number of MFs, it can also increase the computational time required for real-time applications and increase the number of rules leading to the formation of complex rules. The input and output MFs relative to the edge lengths of the triangular structure we use to design this system is shown in Fig. 4.55.

The developed Fuzzy logic controller (FLC) for path tracking used two inputs (Δ_X, Δ_Y) and two outputs (LWS, RWS). Gauss function is preferred for the structure of membership functions. The MFs given above are arranged for functions that take the distance value (Δ_x, Δ_y) of the triangle as an input. These values are normalized between 0 and 1 by multiplication of a constant. Since these values range from 0 to 1, the range of values of the input membership function is determined accordingly. Since the value range for the wheel speeds of the robot is determined between 0 and 40, the input value range of the MFs is determined accordingly. The second step of

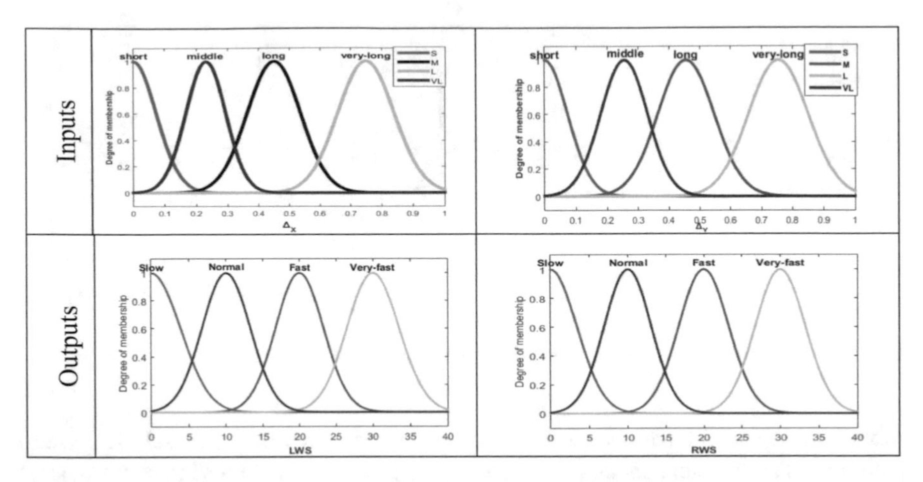

Fig. 4.55 Input/Output membership function for distances (Legends: Δ_X: left distance, Δ_Y: right distance, LWS: left wheel speed, RWS: right wheel speed)

modeling the fuzzy control system is the creation of fuzzy rules. These Rules are based on existing information on the effect of input variables on the output variable. The fuzzy rule base consists of a number of linguistic rule forms ("if…. Then"). The Linguistic variables and their corresponding linguistic terms for fuzzy rules performed here are given in Table 4.6.

The fuzzy inference engine generally considers the methodology of the basic logical operators of AND, OR, NOT to combine the input and output units. The fuzzy logic principle is used to match fuzzy input sets (X1 x… x Xp) based on IF-THEN rules that are interpreted as a fuzzy result to the fuzzy set output. The desired behavior is defined by a set of linguistic rules. For instance, a type-1 fuzzy logic with p inputs (x1 ϵ X1… xp ϵ Xp) and one output (y ϵ Y) with M rules have the following form.

$$R : IF\ x_1\ is\ \widetilde{A}\ and/\ or\ x_2\ is\ \widetilde{B}\ THEN\ y\ is\ G$$

Table 4.6 Linguistic variables and their corresponding linguistic terms

	Linguistic variable		Linguistic terms	Abbreviations of term
Inputs	Distances	Left Distance (Δ_X)	Short, Middle, Long, Very-Long	S, M, L, VL
		Right Distance (Δ_Y)	Short, Middle, Long, Very-Long	S, M, L, VL
Output	Wheel Speeds	Left wheel speed (LWS)	Slow, Normal, Fast, Very-Fast	SL, N, F, VF
		Right wheel speed (RWS)	Slow, Normal, Fast, Very-Fast	SL, N, F, VF

In these experiments, we used type-1 fuzzy sets and a minimum t-norm operation. The rules used for the proposed system are presented in Table 4.7.

The meanings of the abbreviations of Table 4.7 are provided in Table 4.6. The surface shape of the fuzzy inference system obtained for the evaluation of the expert system, which is formed considering all conditions, is shown in Fig. 4.56.

In the proposed system, the final output (net value) was obtained using the Central Centroid (COA) technique, which calculates the center of the total area representing the fuzzy output set essentially. The result graphs obtained with the Type1 Fuzzy Logic control are shown in Fig. 4.57.

Table 4.7 Fuzzy inference rules for WMR path tracking (Type-1)

	Inputs		Outputs	
	Δ_X	Δ_Y	LWS	RWS
1	S	S	SL	SL
2	M	S	N	SL
3	L	S	F	SL
4	VL	S	VF	S
5	S	M	SL	N
6	L	M	F	N
7	M	M	N	N
8	L	M	F	N
9	VL	M	VF	N
10	S	L	SL	F
…				
25	M	VL	N	VF
26	L	VL	F	VF
27	VL	VL	VF	VF

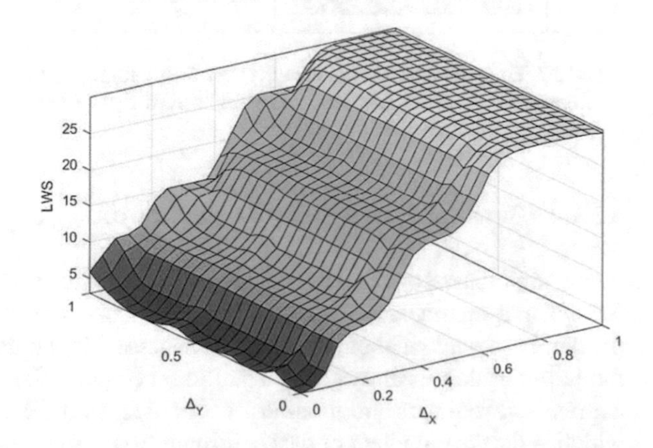

Fig. 4.56 Path tracking control fuzzy surface viewer

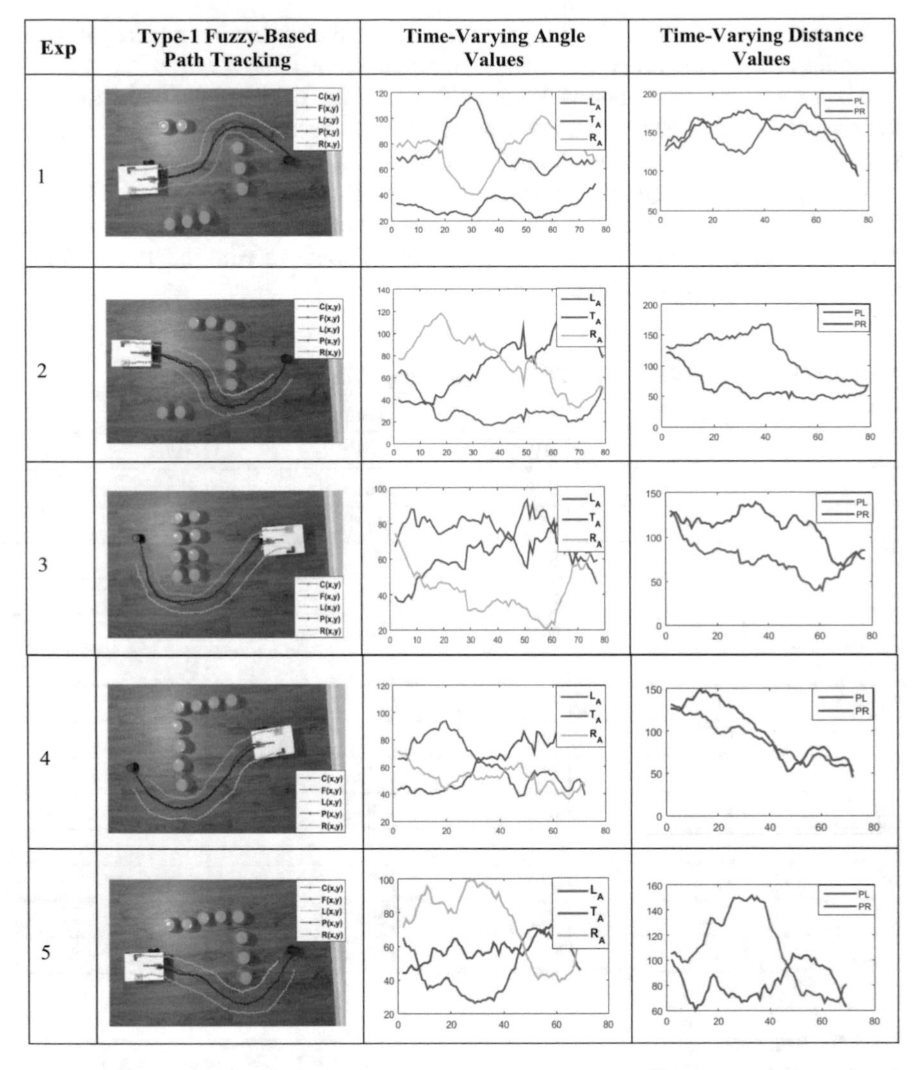

Fig. 4.57 Experimental results using type-1 fuzzy logic-based path tracking algorithm in static environments. (Legend: LA: left wheel angle; RA: right wheel angle; TA: target angle; PL: left distance; PR: right distance)

4.3.4.4 Interval Type 2 Fuzzy Logic Control

Fuzzy logic-based systems are widely used in engineering applications where uncertainties and nonlinear factors are high. Since fuzzy systems can be designed independently of mathematical models, it is successfully applied in controller design and modeling of nonlinear systems. Traditional (type-1) fuzzy sets are not sufficient to express uncertainties and nonlinear properties, so the concept of Type 2 fuzzy sets, which is the development of the traditional fuzzy set type-1, was proposed by Zadeh

[48]. The existence of uncertainties in a nonlinear system control has extended the T1F method to T2F using the highest and lowest values of the parameters.

Type-2 fuzzy clusters are proposed because the projection of uncertainties in input membership functions gives them an extra degree of freedom. One of the essential features of the T2F sets is that they have a third-degree of extra freedom to better express uncertainties in MFs. The structure of the type–2 fuzzy logic system differs from the structure of the type–1 fuzzy logic system in that the type-2 fuzzy logic system has a type reduction mechanism in the output processing block. Therefore, the calculation times are higher because the inference mechanisms of type–2 fuzzy logic systems are more complex than type–1 fuzzy logic systems. For this reason, the researchers proposed a particular type-2 fuzzy logic system called interval type-2 fuzzy systems. Type-2 fuzzy systems, due to the complex internal structures and the high processing load caused by type reduction, type-2 fuzzy systems, a particular case of type-2 fuzzy systems, have been proposed. This is the process of reducing T2F clusters to T1F clusters before defuzzification. Karnik-Mendel (KM) have developed an algorithm for this [53].

Several methods can be used for type reduction (TR) such as centroid type reduction, height type reduction, and center of the set are the most commonly used. In these experiments, a center of sets (cos) type reduction method was used.

When calculating the output of the system with interval type-2 fuzzy clusters, first, the sharp inputs in the fuzzification block are converted to type-2 fuzzy clusters. The defined rules and then the resulting type-2 fuzzy set outputs are converted to type-1 fuzzy set by the type reduction mechanism. The clusters obtained by the type reduction process are converted into crisp outputs by the defuzzification mechanism.

In this section, a different approach from traditional control kinematics is proposed to solve the mobile robot path tracking problem, and a new type-2 fuzzy-based control approach is proposed on this structure. Type 2 fuzzy logic studies in the literature are based on traditional robot control kinematics. In our proposed methods, a triangle-based kinematic control scheme was created and implemented using T2F in our control strategy. This method was used for the first time with type-2 fuzzy logic control. The control application was applied to the real-time experimental environment. Description of type-2 fuzzy (T2F) clusters is given in the path planning stage. The general architecture of the proposed triangular based kinematic structure is given in Fig. 4.50.

In these experiments, we consider the following four inputs: weighted graph-based distances (Δ_x, Δ_y), and triangle shape-based internal angels (R_A, L_A). Since the sum of the internal angles of the triangle is taken into account, the value range (L, R) for the inputs is selected from 0 to 180. These form the range of values of the input element functions. The value range of output member functions varies from 0 to 40. This is the pulse value for mobile robot wheel speeds. The distances from the base to the target coordinate points (Δ_x, Δ_y) are normalized between 0 and 1. Performance analysis was implemented using the triangular based kinematic control diagram, using the edge lengths and internal angles of the triangle. These are used as input parameters of the controller. The main objective is to detect the error that occurs between the virtual path and the real path.

Left and right wheel speeds are outputs that are desired to be obtained against the input values. These output values are important when creating physical functions on the limited region (FOU) that uses the Upper and lower type-1 membership function. The geometry of MFs can be optionally selected. The MFs used in the path tracking using IT2FIS are given in Fig. 4.58.

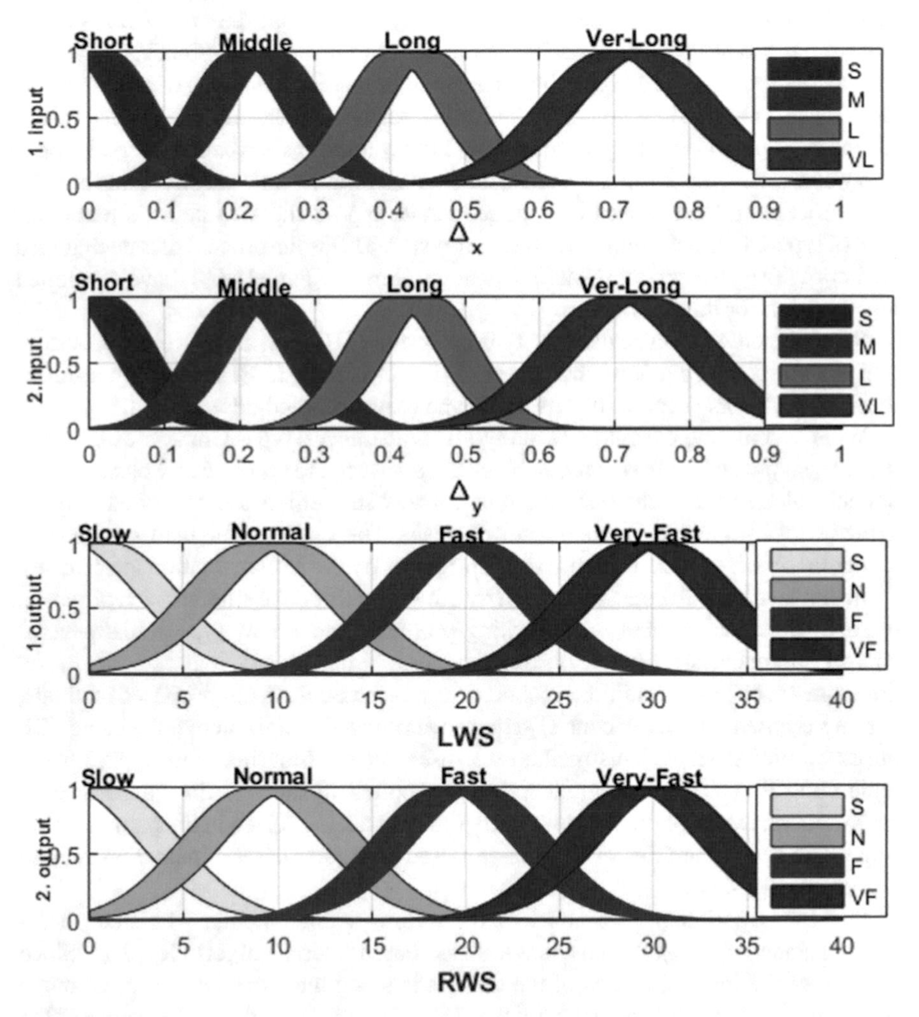

Fig. 4.58 The inputs/outputs membership functions for IT2FIS (Legend: Δ_X: left edge distance; Δ_Y: right edge distance; LWS: left wheel speed; RWS: right wheel speed)

The wheeled mobile robot used in these experiments consists of two servo motors, and each has an incremental encoder. It should be noted that these encoders are not used in this configuration. The wheel speeds are set as the controller output versus the input parameters obtained using the proposed vision-based kinematic control structure given in Fig. 4.50. The robot wheel speeds are controlled using IT2FIS control algorithms to reduce the deviation from the reference path. The critical stage using here is the inference engine, which is an interface that processes input values according to specific rules and produces output type-2 fuzzy sets. A set of linguistic rules defines the desired behavior. For instance, a type-2 fuzzy logic with p inputs (x_1∈X_(1),...x_p∈ X_p), and one output (y∈Y) with M rules have the following form.

$$R^{\ell} : \text{IF } x_1 \text{ is } \widetilde{F}_1^{\ell} \ldots \text{ and } x_p \text{ is } \widetilde{F}_p^{\ell} \text{ THEN } y \text{ is } \widetilde{\mathrm{G}}^{\ell}, \ell = 1 \ldots M$$

Accordingly, the proposed controller, Fuzzy Inference Rule editor for this system, is shown as follows (Table 4.8).

Table 4.8 Fuzzy inference rules for WMR path tracking (Type-2)

	Inputs		Outputs	
	Δ_X	Δ_Y	LWS	RWS
1	S	S	SL	SL
2	M	S	N	SL
3	L	S	F	SL
4	VL	S	VF	S
5	S	M	SL	N
6	L	M	F	N
7	M	M	N	N
8	L	M	F	N
9	VL	M	VF	N
10	S	L	SL	F
...				
25	M	VL	N	VF
26	L	VL	F	VF
27	VL	VL	VF	VF

In these experiments, we used type-2 fuzzy sets and a minimum t-norm operation. The rule firing strength $F^i(x)$ for the crisp input, a vector is given by the type-1 fuzzy set.

$$F^l(x') = \left[\underline{f}^l(x'), \bar{f}^l(x')\right] \equiv \left[\underline{f}^l, \bar{f}^l\right] \tag{4.62}$$

where $\underline{f}^l$ and $\bar{f}^l$ are the lower and upper firing degrees of the l th rule, computed using Eqs. (4.63) and (4.64) in which $*$ represents the t-norm, which is the $prod$ operator in these equations.

$$\underline{f}^l(x') = \underline{\mu}_{\tilde{F}_1^l}(x_1') * \ldots * \underline{\mu}_{\tilde{F}_p^l}(x_p') \tag{4.63}$$

$$\bar{f}^l(x') = \bar{\mu}_{\tilde{F}_1^l}(x_1') * \ldots * \bar{\mu}_{\tilde{F}_1^l}(x_p') \tag{4.64}$$

The fuzzy inference surface constructed by the two inputs and one output parameter surface obtained according to these rules is shown in Fig. 4.59.

Many experimental studies have been performed to evaluate this method. Regarding the obtained results, this algorithm can be easily used on a real robot for path planning based on the results of the experimental study. To carefully test the behavior of the algorithm, numerous experiments on different maps were performed. The experimental results are shown in Fig. 4.60.

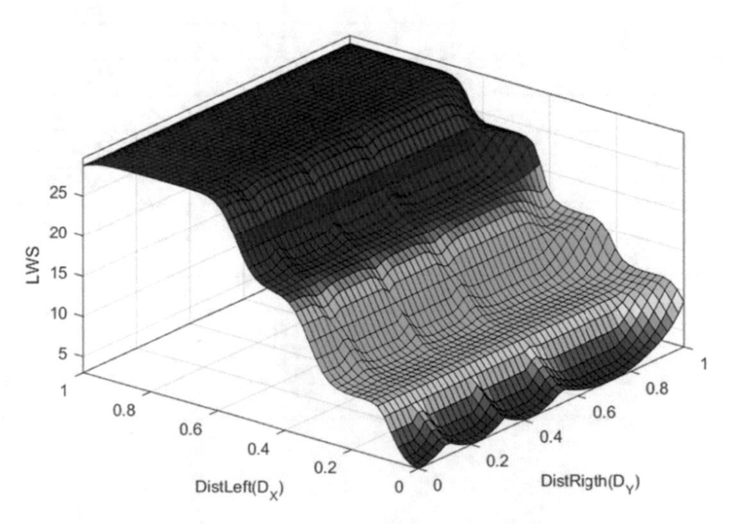

Fig. 4.59 Steering angle control IT2FIS surface viewer

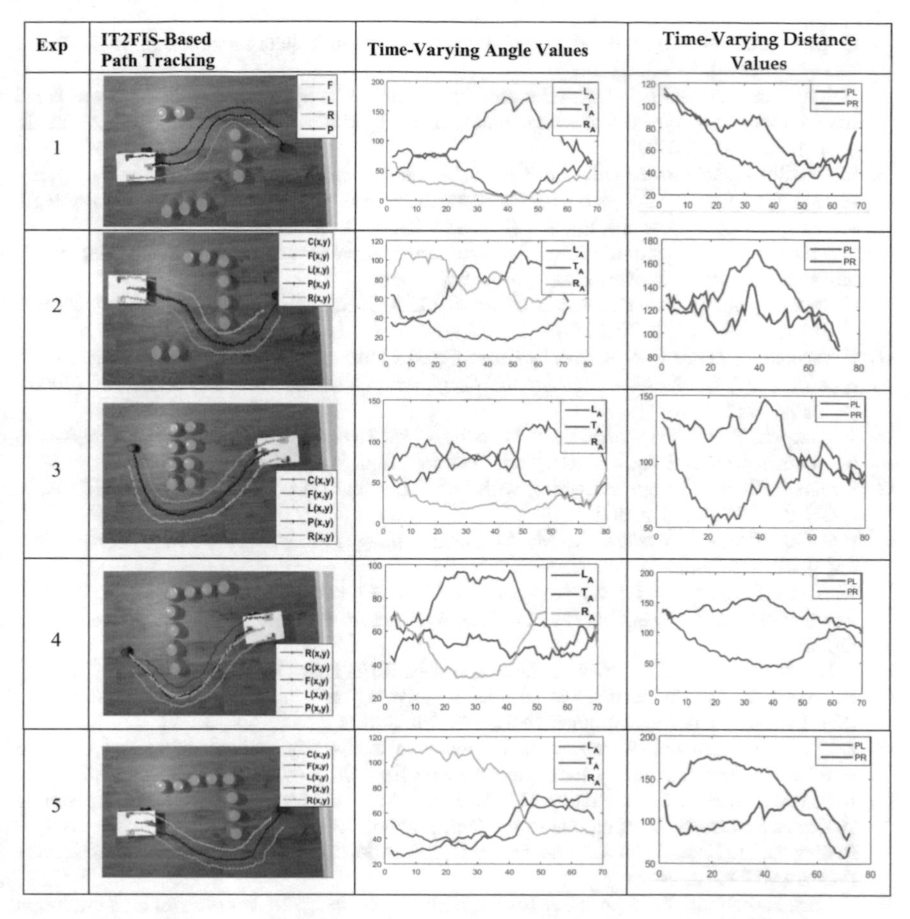

Fig. 4.60 Experimental results using type-2 fuzzy logic-based Path tracking algorithm in static environments. (Legend: LA: left wheel angle; RA: right wheel angle; TA: target angle)

References

1. M. Dirik, A.F. Kocamaz, E. Donmez, Visual servoing based path planning for wheeled mobile robot in obstacle environments, in *International Artificial Intelligence and Data Processing Symposium (IDAP)*, (2017), pp. 1–5
2. M. Dirik, A.F. Kocamaz, E. Donmez, Vision-based decision tree controller design method sensorless application by using angle knowledge, in *Signal Processing and Communication Application Conference (SIU)*, (2016), pp. 1849–1852
3. E. Dönmez, A.F. Kocamaz, M. Dirik, A Vision-based real-time mobile robot controller design based on gaussian function for indoor environment. Arab. J. Sci. Eng. **43**(12), 7127–7142 (2018)
4. N. Baklouti, R. John, A.M. Alimi, Interval type-2 fuzzy logic control of mobile robots. J. Intell. Learn. Syst. Appl. **04**(04), 291–302 (2012)
5. H. Omrane, M.S. Masmoudi, M. Masmoudi, Fuzzy logic based control for autonomous mobile. Comput. Intell. Neurosci. 1–10 (2006)

6. Q. Liang, J.M. Mendel, Interval type-2 fuzzy logic systems: theory and design. IEEE Trans. Fuzzy Syst. **8**(5), 535–550 (2000)
7. R. Martínez, O. Castillo, L.T. Aguilar, Intelligent control for a perturbed autonomous wheeled mobile robot using type-2 fuzzy logic and genetic algorithms. J. Autom. Mob. Robot. Intell. Syst. **1**(2), 12–22 (2008)
8. O. Castillo, L. Amador-Angulo, J.R. Castro, M. Garcia-Valdez, A comparative study of type-1 fuzzy logic systems, interval type-2 fuzzy logic systems and generalized type-2 fuzzy logic systems in control problems. Inf. Sci. (Ny) **354**, 257–274 (2016)
9. W. Fang, L. Zheng, Distortion correction modeling method for zoom lens cameras with bundle adjustment. J. Opt. Soc. Korea **20**(1), 140–149 (2016)
10. C. Manual, *IMAQ Vision Concepts Manual*, 11500 North Mopac Expressw. Austin, Texas 78759-3504 USA, 322916 (2000)
11. F. Yaacoub, Y. Hamam, A. Abche, C. Fares, Convex hull in medical simulations: a new hybrid approach, in *IECON 2006—32nd Annual Conference on IEEE Industrial Electronics*, (2006), pp. 3308–3313
12. F. Yaacoub, Y. Hamam, A. Abche, A 3D convex hull algorithm for modelling medical data in a virtual environment. Int. J. Intell. Syst. Technol. Appl. **5**(1/2), 3 (2008)
13. Thomas Klinger—Image Processing with LabVIEW and IMAQ Vision. Prentice Hall PTR, ISBN: 0-13-047415-0, p. 368
14. R.C. Gonzalez, R.E. Woods, B.R. Masters, Digital Image Processing, 3rd edn. J. Biomed. Opt. **14**(2), 029901 (2009)
15. K. Kwon, S. Ready, *An Introduction with LabVIEW, Practical Guide to Machine Vision Software*, Wiley-VCH Verlag GmbH & Co. KGaA, Boschstr. 12, 69469 Weinheim, Germany All (2006)
16. S. Lowry et al., Visual place recognition: a survey. IEEE Trans. Robot. **32**(1), 1–19 (2016)
17. A. Assa, F. Janabi-Sharifi, Virtual visual servoing for multicamera pose estimation. IEEE/ASME Trans. Mechatronics **20**(2), 789–798 (2015)
18. P.S.B. Divya Agarwal, A review on comparative analysis of path planning and collision avoidance algorithms. Int. J. Mech. Mechatronics Eng. **12**(6) (2018)
19. B.K. Patle, G. Babu L, A. Pandey, D.R.K. Parhi, A. Jagadeesh, A review: on path planning strategies for navigation of mobile robot. Def. Technol. **15**(4) 582–606 (2019)
20. R. Siegwart, I.R. Nourbakhsh, D. Scaramuzza, R.C. Arkin, *Introduction to Autonomous Mobile Robots* (MIT Press, 2011)
21. P. Hart, N. Nilsson, B. Raphael, A formal basis for the heuristic determination of minimum cost paths. IEEE Trans. Syst. Sci. Cybern. **4**(2), 100–107 (1968)
22. L.M.S. Bento, D.R. Boccardo, R.C.S. Machado, F.K. Miyazawa, V.G. Pereira de Sá, J.L. Szwarcfiter, Dijkstra graphs. Discret. Appl. Math. **261**, 52–62 (2019)
23. R. Kala, A. Shukla, R. Tiwari, Fusion of probabilistic A* algorithm and fuzzy inference system for robotic path planning. Artif. Intell. Rev. **33**(4), 307–327 (2010)
24. R. Kala, *On-Road Intelligent Vehicles Motion Planning for Intelligent Transportation Systems British Library Cataloguing-in-Publication Data.* Butterworth-Heinemann, © 2016 Elsevier Inc. (2016)
25. G. Klančar, A. Zdešar, S. Blažič, I. Škrjanc, *Wheeled Mobile Robotics, From Fundamentals Towards Autonomous Systems.* Butterworth-Heinemann, © 2017 Elsevier Inc. (2017)
26. X. Cao, X. Zou, C. Jia, M. Chen, Z. Zeng, RRT-based path planning for an intelligent litchi-picking manipulator. Comput. Electron. Agric. **156**, 105–118 (2019)
27. S.M. LaValle, Rapidly-exploring random trees: a new tool for path planning. Iowa State Univ. Ames, IA 50011 USA **6**(2), 103 (1998)
28. N. Chao, Y. Liu, H. Xia, M. Peng, A. Ayodeji, DL-RRT* algorithm for least dose path Re-planning in dynamic radioactive environments. Nucl. Eng. Technol. **51**(3), 825–836 (2019)
29. N. Chao, Y. Liu, H. Xia, A. Ayodeji, L. Bai, Grid-based RRT∗ for minimum dose walking path-planning in complex radioactive environments. Ann. Nucl. Energy **115**, 73–82 (2018)
30. L. Jaillet, A. Yershova, S.M. La Valle, T. Simeon, Adaptive tuning of the sampling domain for dynamic-domain RRTs, in *IEEE/RSJ International Conference on Intelligent Robots and Systems* (2005), pp. 2851–2856

31. A. Viseras, D. Shutin, L. Merino, Robotic active information gathering for spatial field reconstruction with rapidly-exploring random trees and online learning of gaussian processes. Jr. Sensors **19**(5), 10–16 (2019)
32. E. Donmez, A.F. Kocamaz, M. Dirik, Bi-RRT path extraction and curve fitting smooth with visual based configuration space mapping, in *International Artificial Intelligence and Data Processing Symposium (IDAP)* (2017), pp. 1–5
33. M. Baumann, S. Léonard, E.A. Croft, J.J. Little, Path planning for improved visibility using a probabilistic road map. IEEE Trans. Robot. **26**(1), 195–200 (2010)
34. K. Mrudul, R.K. Mandava, P.R. Vundavilli, An efficient path planning algorithm for biped robot using fast marching method. Procedia Comput. Sci. **133**, 116–123 (2018)
35. P. Wang, S. Gao, L. Li, B. Sun, S. Cheng, Obstacle avoidance path planning design for autonomous driving vehicles based on an improved artificial potential field algorithm. Jr. Energies **12**(12), 2342 (2019)
36. E. Dönmez, A.F. Kocamaz, Design of mobile robot control infrastructure based on decision trees and adaptive potential area methods. Iran. J. Sci. Technol. Trans. Electr. Eng. 1–18 (2019)
37. J. Sun, G. Liu, G. Tian, J. Zhang, Smart obstacle avoidance using a danger index for a dynamic environment. Appl. Sci. **9**(8), 1589 (2019)
38. C.E. Taylor, Adaptation in natural and artificial systems: an introductory analysis with applications to biology, control, and artificial intelligence. complex adaptive systems. John H. Holland. Q. Rev. Biol. **69**(1), 88–89 (1994)
39. A. Basiri, M.A. Oskoei, A. Basiri, A.M. Shahri, Improving robot navigation and obstacle avoidance using kinect 2.0. ICRoM, 486–489 (2017)
40. W.C. Chang, C.H. Wu, Map-based navigation and control of mobile robots with surveillance cameras. Int. J. Adv. Mechatron. Syst. **7**(1), 1 (2016)
41. R. Kala, A. Shukla, R. Tiwari, Dynamic environment robot path planning using hierarchical evolutionary algorithms. Cybern. Syst. **41**(6), 435–454 (2010)
42. L.A. Zadeh, Fuzzy logic, neural networks, and soft computing. Commun. ACM **37**(3), 77–84 (1994)
43. L.A. Zadeh, Soft computing and fuzzy logic. IEEE Softw. **11**(6), 48–56 (1994)
44. M. Dirik, Collision-free mobile robot navigation using fuzzy logic approach. Int. J. Comput. Appl. **179**(9), 33–39 (2018)
45. O. Castillo, Interval type-2 fuzzy logic for hybrid intelligent control. Stud. Fuzziness Soft Comput. **298**, 91–94 (2013)
46. A. Meléndez, O. Castillo, F. Valdez, J. Soria, M. Garcia, Optimal design of the fuzzy navigation system for a mobile robot using evolutionary algorithms. Int. J. Adv. Robot. Syst. **10**(2), 139 (2013)
47. R. Martínez, O. Castillo, L.T. Aguilar, Optimization of interval type-2 fuzzy logic controllers for a perturbed autonomous wheeled mobile robot using genetic algorithms. Inf. Sci. (Ny) **179**(13), 2158–2174 (2009)
48. L.A. Zadeh, The concept of a linguistic variable and its application to approximate reasoning-III. Inf. Sci. (Ny) **9**(1), 43–80 (1975)
49. J.M. Mendel, R.I.B. John, Type-2 fuzzy sets made simple. IEEE Trans. Fuzzy Syst. **10**(2), 117–127 (2002)
50. S.K. Kashyap, IR and color image fusion using interval type 2 fuzzy logic system, in *International Conference on Cognitive Computing and Information Processing(CCIP)*, (2015) pp. 1–4
51. J.M. Mendel, Advances in type-2 fuzzy sets and systems. Inf. Sci. (Ny) **177**(1), 84–110 (2007)
52. O. Castillo, P. Melin, Hybrid intelligent systems for time series prediction using neural networks, fuzzy logic, and fractal theory. IEEE Trans. Neural Netw. **13**(6), 1395–1408 (2002)
53. N.N. Karnik, J.M. Mendel, Q. Liang, Type-2 fuzzy logic systems. IEEE Trans. Fuzzy Syst. **7**(6), 16 (1999)
54. N.N. Karnik, J.M. Mendel, Centroid of a type-2 fuzzy set. Inf. Sci. (Ny) **132**(1–4), 195–220 (2001)

55. M. Dirik, O. Castillo, A. Kocamaz, *Visual-Servoing Based Global Path Planning Using Interval Type-2 Fuzzy Logic Control*, Axioms 2019, Vol. 8, p. 58 (2019)
56. M. Dirik, A.F. Kocamaz, E. Dönmez, *Vision-based decision tree controller design method sensorless application by using angle knowledge*. SIU, 1849–1852 (2016)
57. V. Lippiello, B. Siciliano, L. Villani, Eye-in-hand/eye-to-hand multi-camera visual servoing, in *IEEE Conference on Decision and Control*, (2005), pp. 5354–5359
58. A. Pandey, Mobile robot navigation in static and dynamic environments using various soft computing techniques, Dr. Philos. Dep. Mech. Eng. Natl. Inst. Technol. Rourkela 226 (2016)
59. A. Garcia-Cerezo et al., Using LEGO robots with LabVIEW for a summer school on mechatronics, in *2009 IEEE International Conference on Mechatronics*, pp. 1–6 (2009)
60. R. Grandi, R. Falconi, C. Melchiorri, Robotic competitions: teaching robotics and real-time programming with LEGO mindstorms, in *IFAC Proceedings Volumes (IFAC-PapersOnline)*, (2014), pp. 10598–10603
61. C.C. Robusto, The cosine-haversine formula. Am. Math. Mon. **64**(1), 38 (1957)

Chapter 5
Implementation and Evaluation of the Controllers

To test the proposed methods in a working environment, we created a LabVIEW simulation environment. This simulation environment was used for testing our proposed methods, and real-time applications realized. This would ensure that the algorithm could be easily developed on a real robot on the purpose of path planning and path tracking. All experiments were performed on a 2.40 GHz dual-core system with 16 GB RAM using a combination of LabVIEW and MATLAB.

The initial graph was taken as an input in the form of an image from another device. The black regions are depicted as obstacles, and the white regions are depicted as free space. The distance between the initial and target positions was set and normalized between 0 and 1 for the experiments.

The initial major task was the optimization of global path planning. For this purpose, some parameters were compared. These are criteria such as path length, execution time, and total turning points. The system was tested on several maps with various obstacles. The optimized model was then used for path tracking. These are given in the following subsections.

5.1 Experiments and Performance Analysis of Path Planning Algorithms

Up to this stage, the path planning algorithms and their applications have been emphasized. In this section, a set of benchmark criteria will be presented to determine the preferable one among these algorithms. Next, the path coordinates of the selected algorithm will be used, and real-time applications will be performed. The kinematic control approach and mobile robot applications to be proposed here will be considered and analyzed. The criteria to be considered when determining the optimal path for a mobile robot are measured by various factors such as path length, collision-free area, the total number of turns, and execution time. The purpose of the applications

M. Dirik et al., *Vision-Based Mobile Robot Control and Path Planning Algorithms in Obstacle Environments Using Type-2 Fuzzy Logic*, Studies in Fuzziness and Soft Computing 407, https://doi.org/10.1007/978-3-030-69247-6_5

performed in the first step of this study is to review these planners and compare their performance based on these factors. The experiments have been performed in cluttered and complex unstructured environments with obstacles of different shapes. In this section, the results of all algorithms provided in one place for comparing performance parameters in path length, execution time, and the number of graphs. The following metrics are used to evaluate the performance of algorithms for known environments [1].

- *Path length*: Path Length shows the length of the path taken between the start and the target while performing the robot's task. It determines the total working time required and the total energy consumption of the mobile robot. Therefore, if the path planning algorithm creates the shortest path, it is considered optimal, thus ensuring energy efficiency. That is why it is an essential parameter for real-world solutions.
- *Execution Time*: Describes the time required for application resolution for real-world applications. It is an important criterion, particularly in path planning applications.
- *Total Number of Turns*: The total number of vertices visited while performing the given task is directly related to the memory requirement. Further, the total number of corners, namely turning points, visited during the execution of the task becomes essential.

Taking the explained criterion into consideration, the implementation of the application on the developed simulation environment will be evaluated. In global path planning, the resulting route will be displayed on the robot motion area or on the robot map image.

In our experimental studies, various experiments have been conducted and the results of eight will be compared. The result of eight different application areas represented by nomenclatures ranging from Map1 to Map5 is illustrated in Fig. 5.1. These graphs show all the results (path coordinates) obtained from the algorithms used.

The graphical representations of all the results obtained according to these criteria are shown in the following figures. Plots of the total number of nodes (vertices) in the final path are shown in Fig. 5.2 for all grid-based approaches. Thus, it was observed that GA requires maximum memory to process more cells than other approaches.

The graphical results of path length and execution time for all approaches are shown in Figs. 5.3 and 5.4, together with the visual comparison. After this graphical representation, we have to evaluate and quantify these numerical values in order.

The results obtained from the experiments carried out within the scope of this work were evaluated with two parameters that are the resulting path cost and execution time to obtain the appropriate path. To compare the methods in a robust approach, both the path costs and time Z-Scores (5.1) has been calculated according to the acquired results.

The Z-score indicates the number of standard deviations that a data set leaves above or below the average. The Z-score is also known as the standard score and can be placed on the normal distribution curve. Z-scores range from -3 to $+3$ standard

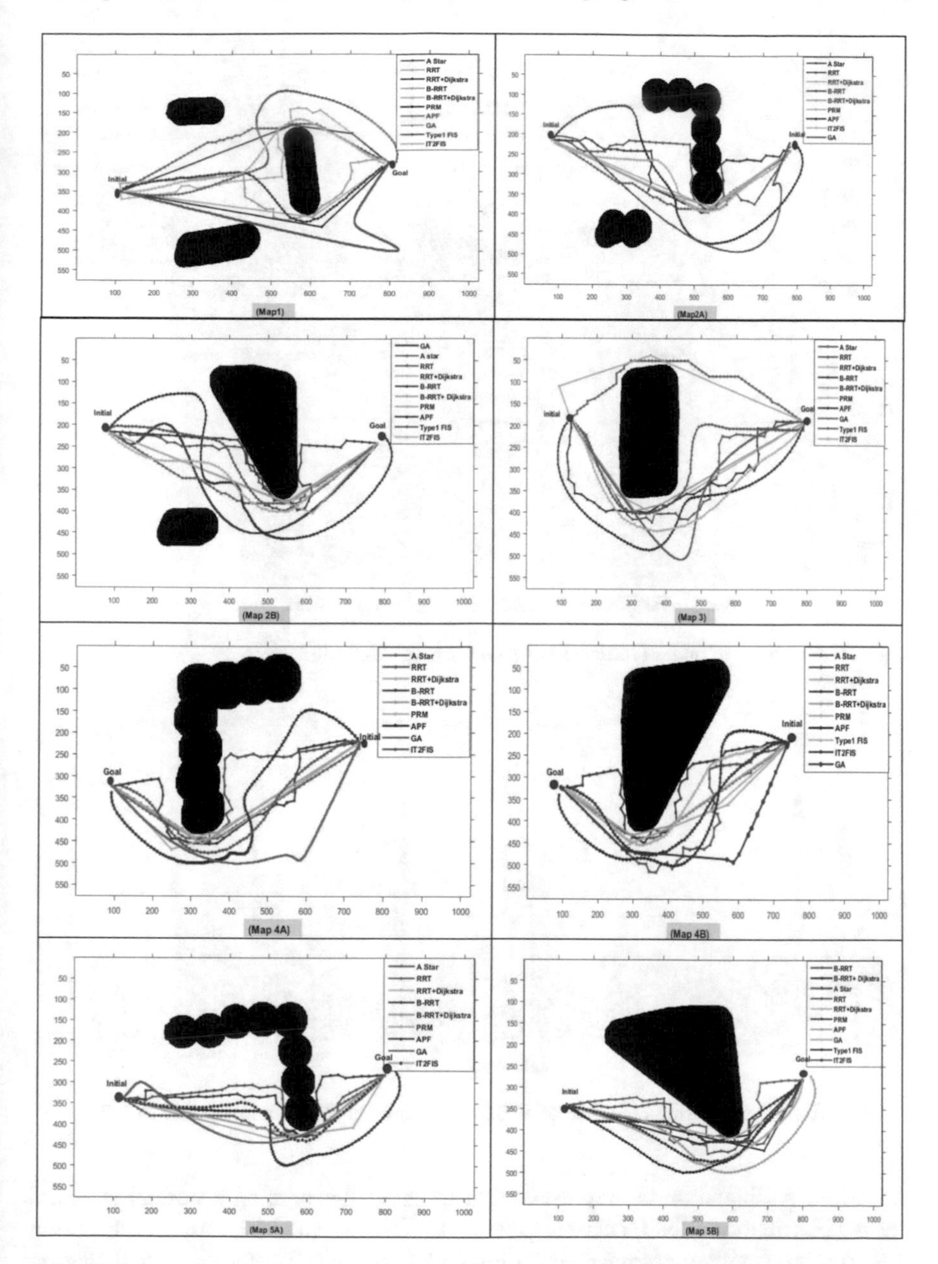

Fig. 5.1 Comparison of planned paths for environment maps Map1 to Map 5B

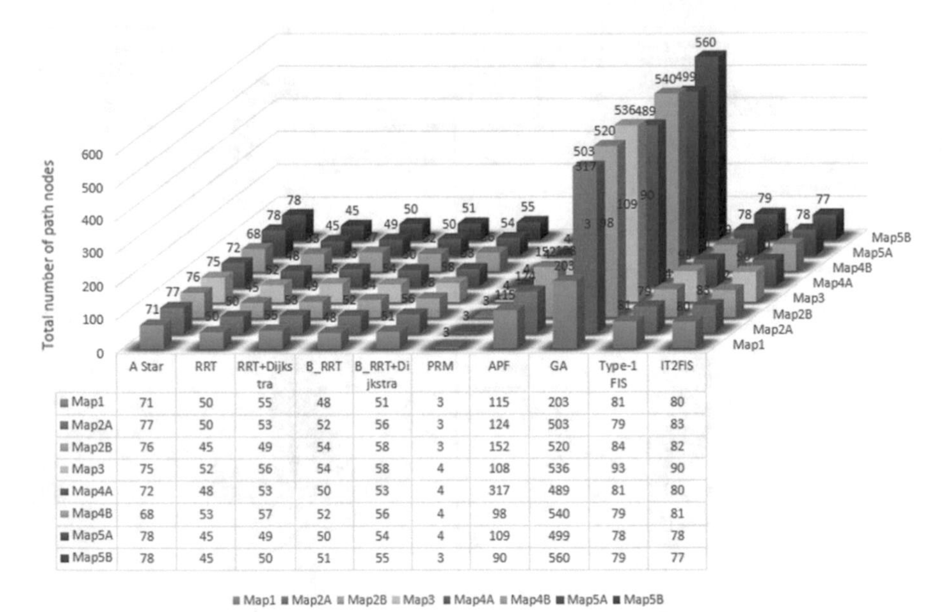

	A Star	RRT	RRT+Dijkstra	B_RRT	B_RRT+Dijkstra	PRM	APF	GA	Type-1 FIS	IT2FIS
Map1	71	50	55	48	51	3	115	203	81	80
Map2A	77	50	53	52	56	3	124	503	79	83
Map2B	76	45	49	54	58	3	152	520	84	82
Map3	75	52	56	54	58	4	108	536	93	90
Map4A	72	48	53	50	53	4	317	489	81	80
Map4B	68	53	57	52	56	4	98	540	79	81
Map5A	78	45	49	50	54	4	109	499	78	78
Map5B	78	45	50	51	55	3	90	560	79	77

Fig. 5.2 A number of nodes (vertices) in a path for maps *Map1* to *Map5B*

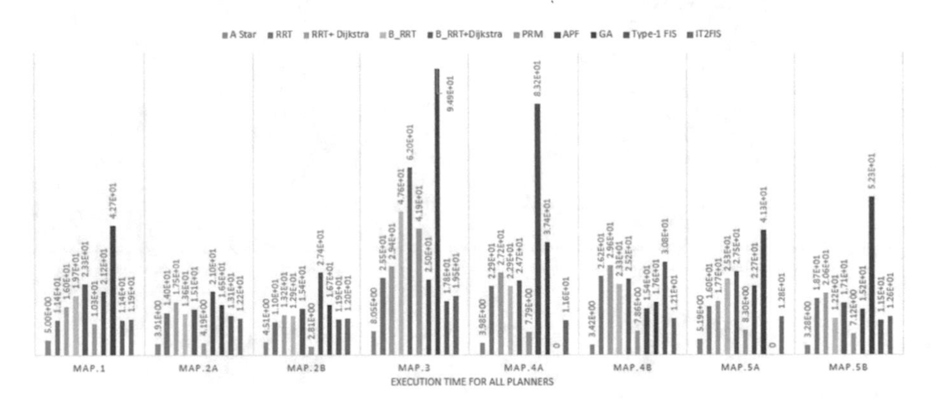

Fig. 5.3 A plot of computational time of all planners for all environment maps M1 to M5

deviation. A diagrammatic representation to show Z-scores on a normal standard normal distribution (SND) curve is given in Fig. 5.5. A standard normal distribution (SND) It is a standard normal distribution with a mean of 0 and a standard deviation of 1.

$$z = \frac{x - \mu}{\sigma} \tag{5.1}$$

In this equation, x represents the input data. The μ parameter corresponds to the average of the series. The σ data in the equation is the standard deviation of the series.

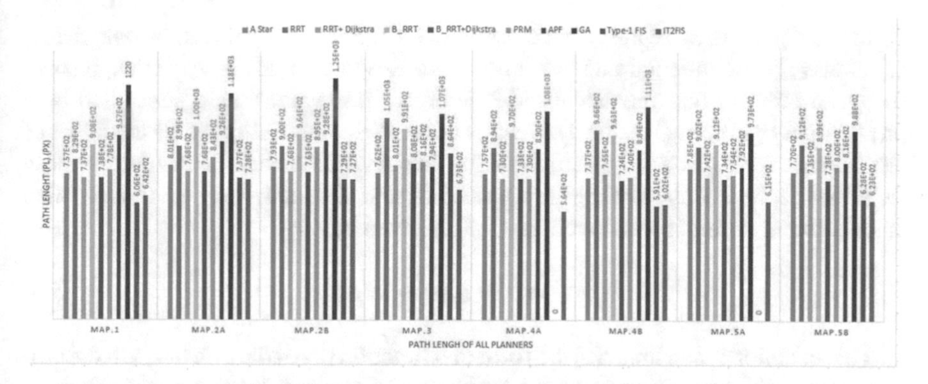

Fig. 5.4 A plot of path lengths of all planners for all environment maps M1 to M5

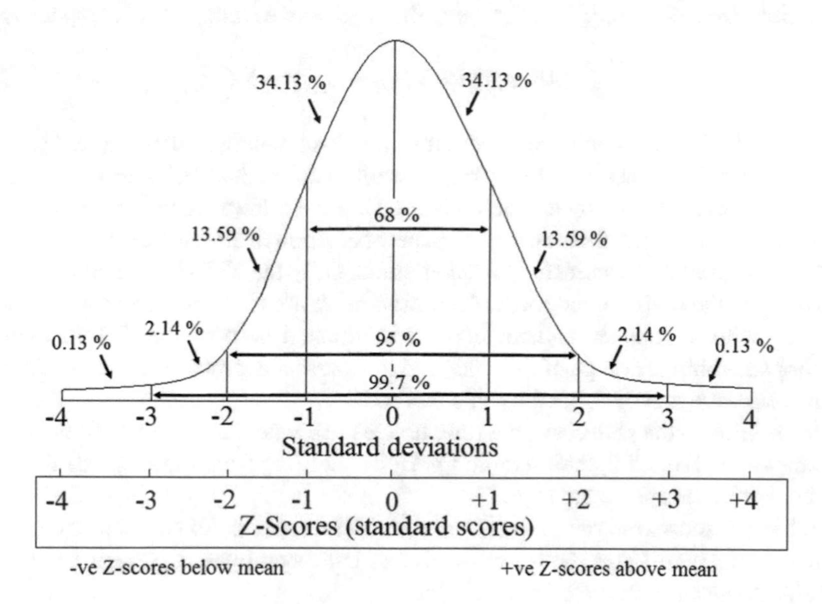

Fig. 5.5 A standard normal distribution (SND)

Before calculating the Z-score, the data must be normalized. This is the method used to standardize the property range of data. Therefore, the calculated Z-Scores were normalized and drawn to the 0–1 range. The following Eq. (5.2) is used for normalization.

$$z_n = \frac{z - z_{min}}{z_{max} - z_{min}} \tag{5.2}$$

In the equation, z_n represents the normalized Z-Score (z) value to be calculated. z_{min} and z_{max} are the minimum and maximum Z-Score values in the series, respectively. In order to fuse the calculated Z-Scores, these values were multiplied by a pre-defined significance factor. The results of the Z-Scores multiplied by the factor of significance were given as input to the value calculation function in the multipurpose functions [2, 3], and the final performance data was obtained. The equation of value calculation in multipurpose functions (5.3) is given below.

$$f(p,t) = sf_i(z_n p) + sf_j(z_n t) \tag{5.3}$$

This equation is calculated in the form of a function depending on the path cost (p) and elapsed time (t). The sf_i and sf_j parameters represent the significance factors; their totals are always equal to 1 (5.4). The $z_n p$ and $z_n t$ parameters in the equation are the normalized Z-Score values for the path length and execution time, respectively.

$$\exists\left(sf_i, sf_j\right) \Rightarrow \forall\left(sf_i + sf_j\right) = 1 \tag{5.4}$$

All methods used in this work were tested in each configuration space. The tests were repeated five times, and the averaged results were evaluated. Table 5.1 shows the path plan costs obtained by the methods used for the different configuration spaces.

Table 5.1 indicated that most of the methods applied in most of the configuration fields have been successfully implemented. Only the T1F method collided with obstacles in the Map.4A and Map.5A configuration areas. Such problems have been handled with the convex hull method, as mentioned before. The IT2FIS method developed within the scope of the study did not cause any problems, and the shortest path costs were mostly formed by this method.

In addition to the path costs, the time it takes to create these paths is an important parameter. In Table 5.2, the execution periods resulting from the methods used for the configuration spaces are given.

Table 5.2 shows the run time of the algorithms applied. To compare the performance of the algorithms, both execution and path length were evaluated together. Details are given in the calculations below.

It should be pointed out that these working periods are the times taken for path planning on the acquired image of the real environment. These times refer to the methods of path formation. In this respect, it can be said that path cost is a more critical parameter. Because it is known that the robot to be operated in a real environment will spend time and energy according to the obtained path costs. In this case, it is stated that the path cost parameter is more critical in terms of enabling the robot to operate efficiently.

Table 5.3 gives the Z-Score values calculated against the path cost values. Z-Score values were not calculated for the methods where the collision actualized, and Z-Score of this related section was accepted as zero ('0').

The table (12) shows the Z-Score values calculated against elapsed time duration values. Z-Score values were not calculated for the method where collision actualized and Z-Score of this related section was accepted as zero ('0') (Table 5.4).

Table 5.1 Path Lengths (PL) obtained in different configuration spaces

Maps	A Star	RRT	RRT + Dijkstra	B_RRT	B_RRT + Dijkstra	PRM	APF	GA	T1F	IT2FIS
Map.1	757.00	829.24	736.78	907.60	737.85	777.83	956.87	1220.0	606.17	641.79
Map.2A	800.56	898.68	768.07	1003.4	768.13	843.28	926.13	1175.0	737.25	727.56
Map.2B	793.29	899.95	767.99	963.75	763.27	894.61	927.89	1254.0	728.99	727.42
Map.3	762.02	1048.7	801.29	990.78	808.41	816.29	794.30	1068.00	864.02	672.83
Map.4A	756.91	893.76	730.30	969.88	732.57	730.25	889.95	1084.00	Collision	564.49
Map.4B	736.88	986.14	754.94	963.29	724.22	740.29	884.17	1111.00	590.91	601.69
Map.5A	785.20	902.20	741.68	912.36	733.60	753.89	792.19	973.00	Collision	615.32
Map.5B	770.20	912.45	735.05	898.74	727.85	799.64	816.26	988.00	628.01	622.85

Table 5.2 Execution Time (ET-sc) periods obtained in different configuration spaces

Maps	A Star	RRT	RRT + Dijkstra	B_RRT	B_RRT + Dijkstra	PRM	APF	GA	T1F	IT2FIS
Map.1	5.00	11.45	15.95	19.66	23.29	10.33	21.19	42.70	11.41	11.93
Map.2A	3.91	13.95	17.54	13.65	15.09	4.19	21.00	16.54	13.07	12.19
Map.2B	4.51	11.05	13.21	12.89	15.38	2.81	27.43	16.66	11.85	12.04
Map.3	8.05	25.47	29.38	47.58	61.99	41.93	24.97	94.91	17.81	19.53
Map.4A	3.98	22.90	27.21	22.91	24.65	7.79	83.19	37.43	Collision	11.65
Map.4B	3.42	26.18	29.65	23.33	25.21	7.86	15.36	17.62	30.79	12.09
Map.5A	5.19	16.03	17.67	25.33	27.49	8.30	22.70	41.25	Collision	12.77
Map.5B	3.28	18.67	20.59	12.23	17.14	7.12	15.19	52.26	11.54	12.64

Table 5.3 Z-Score values calculated for path cost values

Maps	A Star	RRT	RRT + Dijkstra	B_RRT	B_RRT + Dijkstra	PRM	APF	GA	T1F	IT2FIS
Map.1	0.339	−0.068	0.453	−0.510	0.447	0.221	−0.788	−2.271	1.189	0.988
Map.2A	0.456	−0.240	0.686	−0.984	0.686	0.153	−0.435	−2.201	0.905	0.974
Map.2B	0.494	−0.174	0.653	−0.575	0.682	−0.141	−0.350	−2.394	0.897	0.907
Map.3	0.772	−1.427	0.471	−0.983	0.416	0.356	0.525	−1.575	−0.010	1.456
Map.4A	0.384	−0.492	0.554	−0.979	0.540	0.554	−0.467	−1.709	0.000	1.615
Map.4B	0.424	−1.035	0.319	−0.901	0.498	0.404	−0.438	−1.766	1.279	1.216
Map.5A	0.144	−0.917	0.538	−1.009	0.612	0.428	0.080	−1.559	0.000	1.684
Map.5B	0.165	−1.027	0.460	−0.912	0.520	−0.082	−0.221	−1.659	1.356	1.399
Average	0.397	−0.673	0.517	−0.857	0.550	0.237	−0.262	−1.892	0.702	1.280

Table 5.4 Z-Score values for execution time periods

Maps	A Star	RRT	RRT + Dijkstra	B_RRT	B_RRT + Dijkstra	PRM	APF	GA	T1F	IT2FIS
Map.1	1.165	0.554	0.127	−0.225	−0.569	0.660	−0.370	−2.409	0.558	0.508
Map.2A	1.700	−0.155	−0.818	−0.099	−0.364	1.648	−1.458	−0.634	0.008	0.171
Map.2B	1.228	0.257	−0.063	−0.016	−0.386	1.479	−2.173	−0.575	0.138	0.110
Map.3	1.131	0.454	0.302	−0.405	−0.965	−0.185	0.474	−2.244	0.752	0.685
Map.4A	0.971	0.168	−0.015	0.168	0.094	0.810	−2.391	−0.449	0.000	0.645
Map.4B	1.679	−0.749	−1.120	−0.446	−0.646	1.204	0.405	0.163	−1.241	0.753
Map.5A	1.310	0.327	0.179	−0.516	−0.712	1.029	−0.278	−1.961	0.000	0.623
Map.5B	1.028	−0.119	−0.263	0.360	−0.006	0.741	0.140	−2.624	0.412	0.330
Average	1.277	0.092	−0.209	−0.147	−0.444	0.923	−0.706	−1.341	0.078	0.478

Table 5.5 Normalized Z-Score values

NZS	A Star	RRT	RRT + Dijkstra	B_RRT	B_RRT + Dijkstra	PRM	APF	GA	T1F	IT2FIS
NPZS	0.722	0.384	0.759	0.326	0.770	0.671	0.514	0.000	0.602	1.000
NTZS	1.000	0.548	0.433	0.456	0.343	0.865	0.243	0.000	0.542	0.695

: Normalized Z-Score   : Normalized Path Z-Score   : Normalized Time Z-Score

The mean Z-Score values for each method and time obtained from these tables were normalized and drawn to the range 0–1. Normalized Z-Score values are given in Table 5.5.

These Z-Score values calculated for path cost according to the significance factor are given in Table 5.6 that was obtained by multiplying each Z-Score by five different predefined significance factors.

The calculated values of the obtained Z-Score values for the time passing according to the significance factor are given in Table 5.7. This table is obtained by multiplying each Z-Score by five different predefined significance factors.

The mean values of the normalized path length and execution time Z-Scores, which were normalized according to the significance factor, are given in Table 5.8 that expresses the Z-Scores of the fused path and time calculated according to the significant factor.

It was understood from the tables data, and the calculations that IT2FIS method gave the highest Z-Score values when the importance factor of the road cost was higher than 0.5. A higher Z-Score indicates that the measured parametric value is better than the others. It is seen that the A-Star method gives better results if the route cost Z-Score falls below 0.5. As previously methods, it was calculated that the path plan execution times in the configuration spaces obtained from the real working environment. However, it does not have a direct effect on the time it takes to reach the target in the working environment of the actual robot.

On the other hand, it should be stated that the calculated path cost values directly affect the consumed time that the real robot reaches the target in the working environment. Therefore, the IT2FIS method developed within the scope of this work is superior to the other methods in terms of actual working time. In this context, it was observed that the method developed for both path cost and actual runtime parameters yielded compelling results. The best algorithm obtained here will be used in the next step, i.e., the robot path tracking step. The path that this algorithm finds will be the path that the robot will follow in real-time.

Table 5.6 Mean path Z-Score values normalized according to significance factors

Sig. factor	A Star	RRT	RRT + Dijkstra	B_RRT	B_RRT + Dijkstra	PRM	APF	GA	T1F	IT2FIS
SF: 0.80	0.577	0.308	0.607	0.261	0.616	0.537	0.411	0.000	0.482	0.800
SF: 0.65	0.469	0.250	0.494	0.212	0.500	0.436	0.334	0.000	0.391	0.650
SF: 0.50	0.361	0.192	0.380	0.163	0.385	0.336	0.257	0.000	0.301	0.500
SF: 0.35	0.253	0.135	0.266	0.114	0.269	0.235	0.180	0.000	0.211	0.350
SF: 0.20	0.144	0.077	0.152	0.065	0.154	0.134	0.103	0.000	0.120	0.200

Table 5.7 Mean-time Z-Score values normalized by significance factors

Sig. factor	A Star	RRT	RRT + Dijkstra	B_RRT	B_RRT + Dijkstra	PRM	APF	GA	T1F	IT2FIS
SF: 0.20	0.200	0.110	0.087	0.091	0.069	0.173	0.049	0.000	0.108	0.139
SF: 0.35	0.350	0.192	0.151	0.160	0.120	0.303	0.085	0.000	0.190	0.243
SF: 0.50	0.500	0.274	0.216	0.228	0.171	0.433	0.121	0.000	0.271	0.348
SF: 0.65	0.650	0.356	0.281	0.296	0.223	0.562	0.158	0.000	0.352	0.452
SF: 0.80	0.800	0.438	0.346	0.365	0.274	0.692	0.194	0.000	0.434	0.556

Table 5.8 Sum of normalized mean PL and ET Z-Score values according to significance factors

TNZS	A Star	RRT	RRT + Dijkstra	B_RRT	B_RRT + Dijkstra	PRM	APF	GA	T1	IT2FIS
0.80–0.20	0.777	0.417	0.694	0.352	0.684	0.710	0.460	0.000	0.590	0.967
0.65–0.35	0.819	0.442	0.645	0.372	0.620	0.739	0.419	0.000	0.581	0.943
0.50–0.50	0.861	0.466	0.596	0.391	0.556	0.768	0.378	0.000	0.572	0.848
0.35–0.65	0.903	0.490	0.547	0.411	0.492	0.797	0.338	0.000	0.563	0.894
0.20–0.80	0.944	0.515	0.498	0.430	0.428	0.826	0.297	0.000	0.554	0.870

: Total Normalized Z-Score : Best : Second : Third

5.2 Path Tracking Experiments Comparisons

Path tracking is the process of determining speed and steering settings for the robot to follow a specified path. A path consists of positional coordinates of the route, which consists of a particular set of points. The path planning algorithms record the path coordinates when planning and the traced coordinates are followed by the path tracking control algorithms. In mobile robot control applications, traditional robot control kinematics is used. Such control approaches, in particular, path control approaches, are likely to cause systematic or non-systematic errors which are undesirable in control kinematics.

This work aims at overcoming these problems and presents a different perspective by developing new control kinematics developed in our previous studies. Also, four path tracking algorithms are proposed in this work. Two of these were used in our previous studies, and the other two algorithms were used for the first time in the proposed control kinematics. These are fuzzy-based Type-1 and Type-2 soft computing control algorithms. The proposed system includes a combination of soft computing based on visual information processing and a triangular based kinematic control structure. Using this method allows alternative control for more comprehensive and cost-effective indoor navigation.

Experiments were carried out in a static environment and five different configuration environments were utilized. The performances of the proposed algorithms were measured according to the experimental results obtained from real environments. It is observed that the proposed algorithms (Type-1 fuzzy logic, Type-2 Fuzzy Logic, Decision Tree, and Gaussian controller) were able to find the feasible solutions in all test cases.

In the experimental studies, it is aimed to minimize the error rate between robot and virtual paths. The proposed algorithms have been executed on the developed kinematic control structure for this purpose. Several metric values were taken into consideration to measure the performance of the system. In these experiments, the tasks performed with the least error and statistical results of performance metrics in terms of error rate are shown. The statistical analysis and details of these experiments are explained in the following sections.

5.2.1 *Calculation of Path-Cost Between Implementation and Simulation*

To find the difference between the simulation path and the actual path that the robot follows on the way to the target, Eq. (5.5) is used to obtain the ratio of the cost of the path between the expected path and the actual path generated by the movement of the robot until it reaches the target position. The value obtained by the equation shows how much difference occurs between these paths.

$$\frac{D_S}{D_A} = \frac{\sqrt{\sum_{i=1}^{n}(q_i - p_i)^2}}{\sum_{x_1 y_1}^{x_n y_n}\sqrt{\sum_{j=1}^{m}\left(q_j - p_j\right)^2}} \tag{5.5}$$

In this equation, D_S denotes the distance of the simulation path acquired from the path planning process, and D_A denotes the distance of the actual path shaped by Mobile Robot.

5.2.2 *Performance Comparison with Angle Inputs*

Three metric methods have been considered to measure and compare the performance of the proposed controller results in terms of angle input. These metric calculations relate to the error between the center's coordinates of the labels on the robot and the virtual path. Let us illustrate these calculations with a sample (Table 5.9).

Table 5.9 Sample of experimental result

Lx	Ly	Rx	Ry	Fx	Fy	Px	Py	Error (P-F)
114	299	122	376	215	335	245	377	51.61
124	303	133	372	224	340	255	382	52.20
138	299	150	372	233	344	266	384	51.85
178	295	191	375	262	363	309	395	56.85
…								
202	296	214	382	281	381	330	407	55.47
213	301	223	390	288	390	344	414	60.92
225	306	235	400	297	398	355	425	63.97
277	330	286	431	350	427	402	462	62.68
285	334	295	438	357	433	412	465	63.63
296	340	303	444	366	440	420	472	62.76

Table 5.10 Performance metric comparisons for the controller methods with angle input

			Experiments					
			Exp.1	Exp.2	Exp.3	Exp.4	Exp.5	Exp. avg.
Controller methods	Type-2	STDEV	8.19	12.58	5.59	10.54	9.82	*9.343*
		AVGERR	17.97	46.80	13.21	25.53	62.78	*33.258*
		TTLERR	1204.65	3322.80	1043.78	1761.63	4142.77	*2295.127*
	Type-1	STDEV	32.87	5.62	11.89	20.05	6.74	15.434
		AVGERR	88.61	13.33	25.95	28.12	12.49	33.700
		TTLERR	6645.60	1040.15	1972.23	1994.55	849.65	2500.436
	Decision tree	STDEV	7.73	35.67	24.65	8.37	8.88	17.060
		AVGERR	63.33	46.82	38.45	23.48	30.47	40.510
		TTLERR	3981.82	3136.81	2921.82	1643.76	2102.74	2757.390
	Gaussian	STDEV	17.03	13.99	9.38	18.13	196.96	51.098
		AVGERR	25.08	60.56	23.44	43.99	216.19	73.852
		TTLERR	1780.55	3996.93	1875.03	3079.17	14917.25	5129.786
	Error occurring	Min error	1204.65	1040.15	1043.78	1643.76	849.65	1156.398
		Max error	6645.60	3996.93	2921.82	3079.17	14917.25	6312.154

The L(x, y), R(x, y), and F(x, y) given in Table 5.9 correspond to the coordinates of the label at the top of the robot. P(x, y) indicates the virtual path coordinates. Error (P-F) is the error between robot F(x, y) and P(x, y).

The calculations and the metrics obtained are related to the error obtained from this column. These metrics are standard deviation between Cartesian coordinates (STDEV), average Cartesian coordinate error (AVGERR), and total Cartesian coordinate error (TTLERR). Except for these metric values, min and max error is also computed to evaluate total performance comparison between the utilized and developed controller methods. The statistical performance metric values are acquired for each experiment. The comparison of these metrics is shown in Table 5.10.

The obtained results show that the developed IT2FIS methods demonstrate a significant statistical performance compared to the other controllers. It shows the best performance when the average values of the experiments are taken into account. Average performance metrics for the proposed method are shown with blue text in the Table 5.10. In terms of standard deviation, the proposed method has a prominent consistency. Similarly, in terms of error and total error, it has proper performance consistency. The error occurring is categorized with two sub-metrics, which are minimum (Min) and maximum (Max) error occurring. The Min and Max errors are minimum and maximum TTLERR values in the related experiments.

The overall performance of the controller methods is evaluated using path related performance indicators. There are five experiments considered for making the comparison. For observing the best performance, both the simulations (obtained by path planning methods) and the path lengths obtained from the applications (obtained by controller methods via the physical robot) were calculated. The path related values have been given in Table 5.11.

Table 5.11 provides that the IT2FIS algorithm shows satisfying results in both the simulation and implementation path results while using angle values as controller inputs. The average values of the path results show the general performance in all experiments. It should be noted that not only the path length but also the consistency of the controller method is a crucial factor in robotic systems.

The path value changes have a similar pattern for all the controllers in the experiments. There are small differences in controllers. But the most consistent controller is our controller that was developed in this study. Since it is not fluctuated extremely compare to the other controller methods. The difference rate percentage of simulation and implementation is provided in Table 5.12.

These rate values show how much difference occurs between the simulation and implementation paths while using the same configuration space. According to this table, the proposed method shows a prominent performance over the experiments. These rate values are also graphically demonstrated in Fig. 5.6.

The percentage difference between the simulation and the application is an important criterion for demonstrating the consistency of a controller. The smallness of this value is a positive factor for the controller used since it implies that the controller tries to close the simulated path as much as possible. It is derived from the comparisons that the IT2FIS controller has performed better than the other used controllers. In this context, IT2FIS is the best applicable algorithm among the controllers tested on the proposed kinematic control structure.

5.2.3 Performance Comparison with Distance Inputs

Three metric methods have been considered to measure and compare the performance of the proposed controller results in terms of distance input. These metrics are standard deviation between Cartesian coordinates (STDEV), average Cartesian coordinate error (AVGERR), and total Cartesian coordinate error (TTLERR) as emphasized above. Except for these metric values, min and max error is also calculated to evaluate overall performance between the utilized and developed controller methods. The statistical performance metric values are obtained for each experiment. The comparison of these metrics is demonstrated in Table 5.13.

The acquired results show that the developed IT2FIS methods show a significant statistical performance compared to the other controllers. It demonstrates the best performance when the average values of the experiments are taken into consideration. Average performance metrics for the proposed method are indicated by bold text in Table 5.13. In terms of standard deviation, the proposed method has a promising consistency. Similarly, in terms of error and total error, it has significant performance consistency. The error occurring is categorized with two sub-metrics, which are minimum (Min) and maximum (Max) error occurring. The Min and Max errors are minimum and maximum TTLERR values in the related experiments.

Table 5.11 Implementation and simulation path values according to angle inputs

Cont. method	Exp1 (px)		Exp2 (px)		Exp3 (px)		Exp4 (px)		Exp5 (px)		Exp. avg. (px)	
	Imp	Sim	Imp	Sim	Imp	Sim	Imp	Sim	Imp	Sim	Imp. avg.	Sim. avg.
DEC	613	612	760	612	583	677	452	499	631	517	607.8	583.4
Gauss	554	609	687	690	587	622	533	581	602	528	592.6	606
Type-1	573	593	713	707	596	641	509	536	641	614	598	579.4
IT2FIS	543	577	708	678	585	566	484	446	635	588	599.4	609.8

Table 5.12 The difference rate (%) between Impl. and Simulation (IS) path cost

	Exp1	Exp2	Exp3	Exp4	Exp5	Exp. avg.
	IS. rate (%)	IS. rate (%)	IS. rate (%)	IS. rate (%)	IS. rate (%)	IS. rate (%)
DEC	**0.16**	19.47	16.12	10.40	18.07	12.85
Gauss	9.93	6.44	5.96	9.01	12.29	7.52
Type-1	6.26	**4.03**	6.55	7.85	8.27	**6.59**
Type-2	3.49	4.19	**5.25**	**4.30**	**7.31**	**4.90**

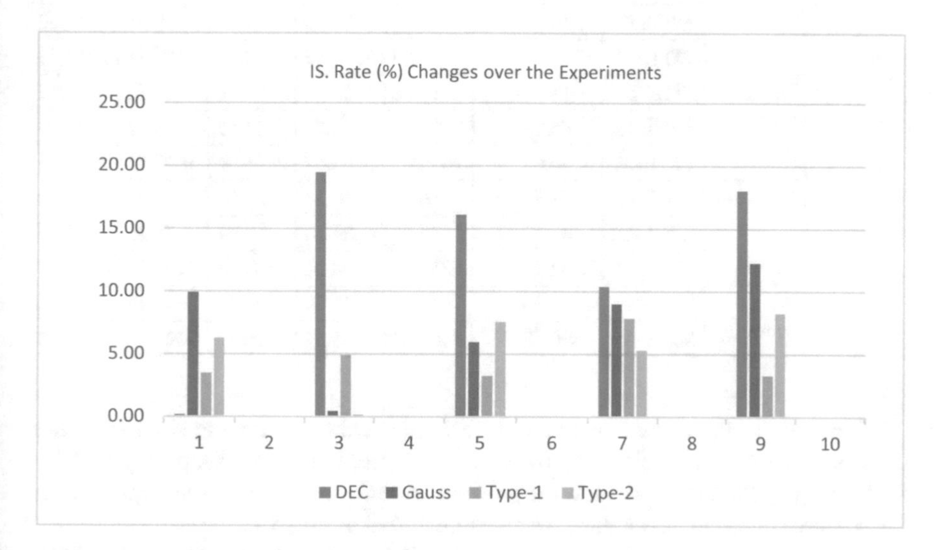

Fig. 5.6 IS. rate (%) changes over the experiments

The general performance of the controller methods is evaluated by using path related performance indicators. There are five experiments that are considered for making the comparison. To observe the controller performances, both the simulations (obtained by path planning methods) and the path lengths obtained from the applications (obtained by controller methods via the physical robot) were calculated—the path related values presented in Table 5.14.

Table 5.14 provides that the IT2FIS algorithm shows satisfying results in both the simulation and implementation path results while using distance values as controller inputs. The average values of the path results show the general performance in all experiments.

The path value changes have a similar pattern for all the controllers in the experiments. There are small differences in controllers. Nevertheless, the most consistent controller is the developed controller within this study. Since it is not fluctuated extremely compare to the other controller methods. The difference between simulation and implementation is given as a percentage in Table 5.15.

Table 5.13 Performance metric comparisons for the controller methods with distance input

			Experiments					
			Exp.1	Exp.2	Exp.3	Exp.4	Exp.5	Exp. avg.
Controller methods	Type-2	STDEV	23.95	7.71	6.79	9.39	8.39	***11.25***
		AVGERR	63.26	23.38	16.29	22.22	13.15	***27.66***
		TTLERR	4744.21	1823.43	1238.00	1577.38	893.98	***1428.30***
	Type-1	STDEV	11.17	20.79	13.83	14.73	15.55	15.21
		AVGERR	26.81	76.87	36.72	27.69	59.91	45.60
		TTLERR	1796.04	5457.69	2900.65	1910.72	4253.44	3263.71
	Decision tree	STDEV	18.39	14.86	31.61	14.48	19.52	19.77
		AVGERR	66.18	24.79	42.70	34.69	55.79	44.83
		TTLERR	4830.83	1636.09	3245.57	2498.04	3849.64	3212.03
	Gaussian	STDEV	7.70	16.83	10.34	25.99	16.09	15.39
		AVGERR	30.32	79.57	21.35	58.67	75.23	53.03
		TTLERR	2152.93	5251.43	1708.01	4106.73	5190.92	4566.35
	Error occurring	Min error	1796.04	1636.09	1238.00	1577.38	893.98	1428.30
		Max error	4830.83	5457.69	3245.57	4106.73	5190.92	4566.35

These rate values show how much difference occurs between the simulation and implementation paths while using the same configuration space. According to Table 5.15, the proposed method shows a prominent performance over the experiments. These rate values are also graphically demonstrated in Fig. 5.7.

The percentage difference between the simulation and the application is an essential criterion for demonstrating the consistency of a controller. The smallness of this value is a decisive factor for the controller used since it implies that the controller tries to close the simulated path as much as possible. It is understood from the comparisons that the IT2FIS controller has performed better than the other used controllers. In this context, IT2FIS is the best applicable algorithm among the controllers tested on the proposed kinematic control structure.

5.2.4 Implementation Comparison for Designed System Inputs

The Summary results of the implementation path lengths have been demonstrated in the following Table 5.16. The proposed methods that use the distance parameter as input have outperformed both designed and previous methods, which use the angle parameter as controller input. The best average path length result has been acquired with the Type-2 controller, which takes the distance parameter as input.

Table 5.14 Implementation and simulation path values according to distance inputs

Cont. method	Exp1 (px)		Exp2 (px)		Exp3 (px)		Exp4 (px)		Exp5 (px)		Exp. avg. (px)	
	Imp	Sim	Imp	Sim	Imp	Sim	Imp	Sim	Imp	Sim	Imp. avg.	Sim. avg.
DEC	588	610	756	710	582	677	421	529	601	**517**	589.6	608.6
Gauss	575	609	686	690	610	622	529	581	606	587	601.2	617.8
Type-1	582	593	729	**678**	591	**566**	443	446	631	614	595.2	579.4
IT2FIS	**524**	**542**	**717**	704	**580**	603	**407**	**442**	**599**	588	**565.6**	**575.8**

Table 5.15 The difference rate (%) between Impl. and Simulation (IS) path cost

	Exp1	Exp2	Exp3	Exp4	Exp5	Exp. avg.
	IS. rate (%)	IS. rate (%)	IS. rate (%)	IS. rate (%)	IS. rate (%)	IS. rate (%)
DEC	3.74	6.08	16.72	25.65	13.98	14.10
Gauss	5.91	**0.58**	1.97	9.83	3.14	4.29
Type-1	**1.89**	7.00	4.23	**0.68**	2.69	**3.30**
Type-2	3.44	1.81	**3.97**	8.56	**1.84**	**3.14**

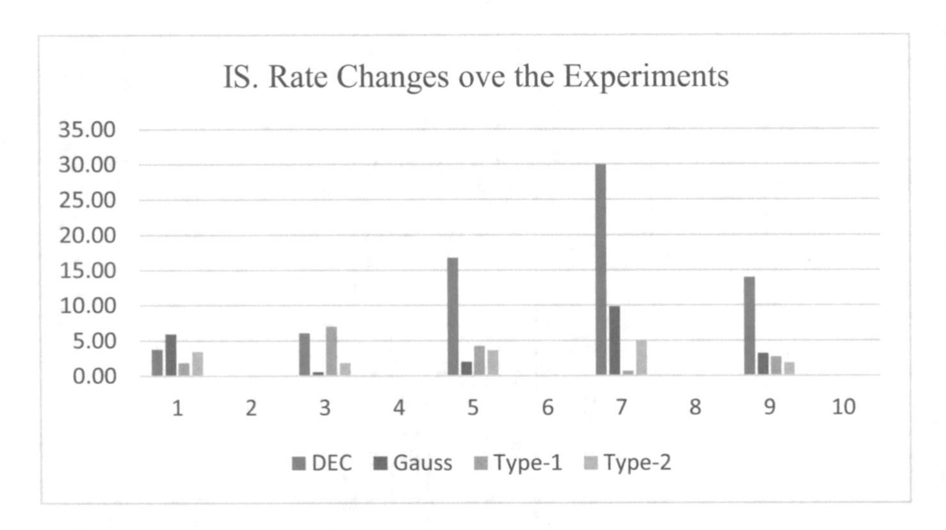

Fig. 5.7 IS. rate (%) changes over the experiments

Table 5.16 General path length results according to the controller inputs

	Exp1 (px)	Exp2 (px)	Exp3 (px)	Exp4 (px)	Exp5 (px)	Exp (px) avg.
DEC-A	613	760	583	452	631	607.8
DEC-D	**588**	**756**	**582**	**421**	**601**	**589.6**
Gauss-A	**554**	687	**587**	533	**602**	**592.6**
Gauss-D	575	**686**	610	**529**	606	601.2
Typ-1-A	**573**	**713**	596	509	641	598
Typ-1-D	582	729	**591**	**443**	**631**	**595.2**
Typ-2-A	543	**708**	585	484	635	599.4
Typ-2-D	**524**	717	**580**	**407**	**599**	**565.6**

Table 5.17 shows the percentage difference between the implementation and simulation path length (or implementation and simulation rate—IS. Rate) results of the conducted experiments. The results show that best IS rate values are acquired with

Table 5.17 The difference rate (%) between Impl. and Simulation (IS) path cost

	Exp1	Exp2	Exp3	Exp4	Exp5	Exp. avg.
	IS. rate (%)	IS. rate (%)	IS. rate (%)	IS. rate (%)	IS. rate (%)	IS. rate (%)
DEC-A	**0.16**	19.47	**16.12**	**10.40**	18.07	**12.85**
DEC-D	3.74	**6.08**	16.72	25.65	**13.98**	14.10
Gauss-A	9.93	6.44	5.96	9.01	12.29	7.52
Gauss-D	**5.91**	**0.58**	**1.97**	9.83	**3.14**	**4.29**
Typ-1-A	6.26	4.03	6.55	7.85	8.27	6.59
Typ-1-D	**1.89**	7.00	**4.23**	**0.68**	**2.69**	**3.30**
Typ-2-A	3.49	4.19	5.25	4.30	7.31	4.90
Typ-2-D	**3.44**	**1.81**	**3.97**	8.56	**1.84**	**3.14**

the proposed Type-1 and Type-2 methods. Besides, the best average IS Rate value according to the input parameters has been acquired with the distance input parameter.

References

1. I. Noreen, A. Khan, K. Asghar, Z. Habib, A path-planning performance comparison of RRT*-AB with MEA* in a 2-dimensional environment. Symmetry (Basel) **11**(7), 945 (2019)
2. Z. Zhu, J. Xiao, J.-Q. Li, F. Wang, Q. Zhang, Global path planning of wheeled robots using multi-objective memetic algorithms. Integr. Comput. Aided Eng. **22**(4), 387–404 (2015)
3. M. Nazarahari, E. Khanmirza, S. Doostie, Multi-objective multi-robot path planning in continuous environment using an enhanced genetic algorithm. Expert Syst. Appl. **115**, 106–120 (2019)

Chapter 6
Conclusion and Future Work

In this book, we proposed a vision-based kinematic control structure for a wheeled mobile robot (WMR) path planning and path tracking problems in a partially cluttered indoor environment using various heuristics and Soft computing algorithms. The aim was to develop a behavioral strategy for mobile robot navigation in static indoor environments. To accomplish this purpose, a hybrid control architecture was proposed to combine the advantage of vision sensors and intelligent control algorithms that are Type-1 fuzzy logic, Type-2 fuzzy logic, Genetic algorithm that can produce a powerful automatic control system. Modeling and controlling the complex non-linear system, the fuzzy logic control method has been applied in various areas. The type-2 fuzzy logic system is superior to the type-1 fuzzy logic system since type-2 fuzzy provides better results with uncertainties and non-linear system controls.

The proposed system for global path planning and tracking problems consists of three stages. First, visual information was obtained by using video frames obtained from an overhead camera. According to this visual information, robot, target, and obstacles position information was obtained. In this stage, the NI Vision Development Module is used to analyze objects (robot, target, obstacles) in the acquired images. This module includes hundreds of functions to process acquired images and provides access to almost all the vision functionality available in LabVIEW. Second, the input parameters of the path planning algorithms were obtained from the images processed in the previous step, and unobstructed path planning has been formed. In total, eight algorithms are used as planning algorithms, and their performances have been compared and analyzed. Regardıng to the results, the IT2FIS algorithm is the best among the algorithms in terms of execution time and the path length. In this stage, it is observed that IT2FIS is superior and outperforms in most of the test environments. For this reason, path coordinates generated by IT2FIS has been utilized in the robot path tracking stage. Last, the third stage is the path-follow phase, where the proposed control structure is used.

The previously developed two algorithms and proposed algorithms allow the robot to follow the specified path. The algorithms for the proposed structures are Type-1

M. Dirik et al., *Vision-Based Mobile Robot Control and Path Planning Algorithms in Obstacle Environments Using Type-2 Fuzzy Logic*, Studies in Fuzziness and Soft Computing 407, https://doi.org/10.1007/978-3-030-69247-6_6

Fuzzy logic, IT2FIS and previously developed controllers are Gaussian, and Decision Tree controllers. These methods were applied to the designed geometric shape-based control structure (Distance-Based Triangle Structure and weighted Graph-Based Controller Design). The proposed system has been successfully applied in various configuration maps with static obstacles, and their performances have been compared. These comparisons are illustrated with tables and graphical representations. These comparisons indicate that the proposed IT2FIS algorithm provides better results when compared to the other algorithms. In addition, regarding the experimental results, the designed Path Planning methods and Controllers are accurate and efficient for robot motion control.

This book includes the following achievements:

- To create a path plan, two fuzzy-based algorithms are developed, and these algorithms have been compared with commonly known path planning methods. The path planning rules created for both Type-1 and Type-2 algorithms.
- The developed two new fuzzy-based controllers characterize the mobile robot behavior. These controllers are compared with the previously developed Gaussian and Decision Tree-based controllers.
- Color-based threshold and template matching are used together to increase the efficiency of the object detection process. In this way, frame loss has been prevented as much as possible.
- A distance-based triangle structure has been developed to generate controller inputs which used to characterize mobile robot motions. This structure is used for the first time in a real application as well.
- LabVIEW software is used to implement, configure, and test the proposed navigation systems and controllers. The LabVIEW is specially configured for robotic applications and provides a high level of programmability.
- Experimental results are evaluated with both path lengths and path forming time. In addition, for the first time, general path tracking performance is compared with statistical approaches and Z-Scores.
- Path plans have been tracked according to the adaptive threshold computation method. The distance between the wheels is used as a first input to trigger the adaptive method.
- In addition to simulation experiments, physical world experiments have been implemented for an eye out device configured robot control system.
- Soft computing algorithms and AI-based methods have been used together to create an efficient robot control framework.

In our future work, we plan to use multi-vision sensor data to minimize uncertainties in the environment by learning more about the state of 3D obstacles by using multiple cameras on a robot traveling in a dynamic environment. By integrating the proposed control structure, we aim to develop robust robotic systems capable of performing physical tasks in the real world.

Index

M. Dirik et al., *Vision-Based Mobile Robot Control and Path Planning Algorithms in Obstacle Environments Using Type-2 Fuzzy Logic*, Studies in Fuzziness and Soft Computing 407, https://doi.org/10.1007/978-3-030-69247-6

Printed in the United States
by Baker & Taylor Publisher Services